Archaeoentomological and Archaeoparasitological Reconstructions at Îlot Hunt (CeEt-110)

New perspectives in historical archaeology (1850-1900)

Allison Bain

BAR International Series 973
2001

Published in 2016 by
BAR Publishing, Oxford

BAR International Series 973

Archaeoentomological and Archaeoparasitological Reconstructions At Îlot Hunt (CeEt-110)

ISBN 978 1 84171 260 4

BAR Publishing is the trading name of British Archaeological Reports (Oxford) Ltd. British Archaeological Reports was first incorporated in 1974 to publish the BAR Series, International and British. In 1992 Hadrian Books Ltd became part of the BAR group. This volume was originally published by Archaeopress in conjunction with British Archaeological Reports (Oxford) Ltd / Hadrian Books Ltd, the Series principal publisher, in 2001. This present volume is published by BAR Publishing, 2016.

Printed in England

BAR
PUBLISHING

BAR titles are available from:

BAR Publishing
122 Banbury Rd, Oxford, OX2 7BP, UK
EMAIL info@barpublishing.com
PHONE +44 (0)1865 310431
FAX +44 (0)1865 316916
www.barpublishing.com

Abstract

The latter half of the nineteenth century was a time of great change in urban North America as sanitary by-laws were passed, permanent boards of health created, and sewer systems and waterworks installed in many cities. Advances in medicine, resulting in the development of germ theory and bacteriology, allowed doctors to better understand the prevention and treatment of the epidemic illnesses that continued to ravage North American cities. Local hygienic movements also encouraged citizens to improve their domestic sanitary practices. However, in contrast, documentary sources from this time period imply that urban settings in North America continued to be squalid and unliveable. It is therefore unclear, due to these conflicting sources, whether North American cities were healthier places to live in the latter half of the nineteenth century.

A multidisciplinary contextual approach, examining both social and biological variables, is used to access a more accurate or realistic view of the past. At the Îlot Hunt site in Québec City (CeEt-110), a mixture of domestic and commercial establishments in the port area, documentary sources and material culture, in the form of sanitary structures and artifacts, were combined with biological data derived from preserved insect and parasite egg remains or archaeoentomology and archaeoparasitology respectively. Archaeoentomological samples from two latrines, a drain, and an abandoned well or cistern structure at the site, resulted in the identification of 6755 insects from 48 levels. However, the archaeoparasitological analysis of a parallel set of samples, yielded very little data, which is either the result of post-depositional factors or the lack of faecal remains within the structures investigated.

The outcome of the environmental analysis, in combination with the material culture data, indicate that the inhabitants of the Îlot Hunt site, adapted to the economic challenges facing them, and maintained a relatively clean yard environment. While it appears they heeded the city's sanitary by-laws, and were probably attached to the city's water and sewer networks, their medical practices went unchanged throughout this time period, regardless of the new approaches promoted by health reformers of the late nineteenth century. Through this multidisciplinary contextual analysis, a new view of a late nineteenth-century Québec City has been accessed, indicating the value of this approach in historical archaeology.

Résumé

La deuxième moitié du XIXe siècle fut une période de grands changements dans les milieux urbains d'Amérique du Nord: des ordonnances sanitaires furent créées, des égouts et aqueducs construits et des bureaux de santé permanents implantés dans plusieurs villes. En médecine, le progrès des études bactériologiques a permis aux médecins de mieux comprendre la prévention et le traitement des épidémies qui continuaient alors de ravager les villes nord-américaines. De plus, des mouvements hygiéniques incitaient les citoyens à améliorer leurs pratiques sanitaires domestiques. Pourtant, les sources écrites de l'époque indiquent que ces milieux urbains étaient encore malsains et insalubres. La réalité sanitaire urbaine de la deuxième moitié du XIXe siècle est donc incertaine puisque les sources sont contradictoires.

Une approche contextuelle multidisciplinaire examinant les traits sociaux et biologiques est utilisée afin d'accéder à une vision plus réaliste du passé. Au site de l'Îlot Hunt à Québec (CeEt-110) un mélange d'établissements domestiques et commerciaux à proximité du port, de sources écrites et de culture matérielle sous forme de structures sanitaires et d'artefacts est confronté aux données biologiques provenant de restes préservés d'insectes (archéoentomologie) et de parasites (archéoparasitologie). Les échantillons archéoentomologiques prélevés dans deux fosses d'aisance, un drain et un puits abandonné ou citerne ont permis, à travers les 48 couches archéologiques, l'identification de 6755 insectes. Par contre, l'analyse archéoparasitologique de ces mêmes échantillons n'a livré que très peu de données. Cette situation résulte soit de la destruction post-dépositionnelle des oeufs ou de l'absence de matière fécale dans les structures étudiées.

Les résultats de l'analyse environnementale, combinés aux restes de la culture matérielle, démontrent que les habitants du site de l'Îlot Hunt se sont adaptés aux défis économiques auxquels ils étaient confrontés et ont su maintenir une certaine salubrité dans leur cour. Alors qu'ils suivaient les ordonnances sanitaires de la ville et qu'ils étaient reliés aux réseaux d'égouts et d'aqueducs, les habitants, indifférents aux nouvelles approches promues par les réformes de la santé de l'époque, n'ont pas modifié leurs pratiques médicales. Par cette analyse contextuelle multidsciplinaire, une nouvelle vision de la ville de Québec au dix-neuvième siècle a été obtenue, montrant ainsi la valeur d'une telle approche en archéologie historique.

TABLE OF CONTENTS

Abstract/Résumé **i**
Table of Contents **ii-iii**
List of Tables **iv**
List of Figures **iv**
Appendix A- Figures **iv**
Appendix B- Tables **v**
Chapter One- Introduction **1-11**
1.1 Introduction to the project 1
1.2 An interdisciplinary analytical approach to studying Îlot Hunt 2
1.3 Archaeoentomology- Current trends 3
Environmental reconstructions 4
Insect biogeography 5
Reconstruction of agricultural practices and storage estimates 5
Health and sanitation 5
Answering archaeological questions 6
1.4 Archaeoparasitology- Current trends 6
Literature review of archaeoparasitological studies 7
Archaeoparasitology in Northeastern North America 8
Parasitological analysis at the household level 8
Resistance to improved sanitation 8
Evidence for chronic parasitism 8
1.5 Studying health and sanitation 9
The documentary record 9
Material culture studies 9
The Îlot Hunt sanitary system 9
1.6 Statement of hypothesis 10
Chapter Two – Urban life in nineteenth-century North America **12-22**
2.1 Life in the nineteenth century city 12
2.2 Nineteenth century medical developments, beliefs and practices 14
2.3 Life in Québec City 1850-1900 16
The municipal aqueduct 17
The installation of sewers and the privy pit menace 18
Street maintenance 20
Sanitary by-laws and the formation of the Conseil d'hygiène de la province de Québec 20
Chapter Three - The Îlot Hunt site **23-37**
3.1 History of the Îlot Hunt site 23
History of Lower Town, Québec City 23
The History of the Îlot Hunt Site (CeEt-110), Lower Town 24
Îlot Hunt investigations 1987-1994 25
3.2 1995 Excavations at Îlot Hunt 26
Sub-operation 16C 26
Sub-operation 16D 28
Sub-operations 16K and 16M 29
Sub-operations 16E and 16F 29
Summary of the depositional sequences within the 16D200/16C26 privy 30
Surveys 16A, 16B, 16G, and 16J 30
3.3 Specialist analyses relevant to the 1995 excavations 30
Socio-economic analysis 30
Archaeozoological analysis 31
Palaeobotanical analysis 31
Summary of Îlot Hunt Levels from 1993 and 1995 excavations 32
3.4 Medical and sanitary practices at Îlot Hunt 33
Material culture and structural remains dating to c. 1850 33
Material culture and structural remains dating between 1860-1875 33
Material culture and structural remains dating from 1880-1900 34
Documentary evidence related to sanitation at Îlot Hunt 34
Demographic data 34
Documented infrastructure data 35
Îlot Hunt 1850-1900: a textual and material view of sanitation and health 35

Chapter Four- Archaeoentomology and archaeoparasitology Methodology **37-41**
4.1 Archaeoentomological methodology 37
Field procedure 37
Laboratory procedure 37
4.2 Archaeoparasitological methodologies 37
Flotation technique 37
Dilution technique 37
Palynological technique 38
Archaeoparasitology methodology for Îlot Hunt analysis 39
Field procedure 39
Laboratory procedure 39
4.3 Pitfalls in archaeoentomological and archaeoparasitological interpretations 39
Pitfalls in archaeoentomological interpretations 39
Pitfalls in archaeoparasitological interpretations 40
Chapter Five- Results of archaeoentomological and archaeoparasitological investigations **42-54**
5.1 The insect and parasite remains recovered from Îlot Hunt 42
Archaeoentomological and archaeoparasitological remains dating to 1825-1875 (16D25) 42
Archaeoentomological remains dating to c. 1850 43
Privy deposit 16C74 dating to c. 1850 43
16E100 structure dating to c. 1850 45
Archaeoentomological and archaeoparasitological remains dating to c. 1860 45
16D200 layers dating to c. 1860 45
Privy deposit 16C70 dating to c. 1860 48
16D203 drain dating to c. 1860 48
Archaeoentomological and archaeoparasitological remains dating to c. 1870 49
Archaeoentomological and archaeoparasitological remains dating to c. 1875 50
Archaeoentomological and archaeoparasitological remains dating from 1880-1900 50
16D200 fill dating from 1880-1900 50
Layers related to late nineteenth-century renovation of the 1875 building 51
Layers from a late 1800s depression below 1875 building 51
5.2 Summary of the Îlot Hunt insect and parasite faunas 52
Summary of the Îlot Hunt archaeoparasitological analysis 53
Chapter Six – Discussion **55-70**
6.1 Background noise and ecological groupings in the Îlot Hunt insect fauna 55
The background fauna 55
Ecological groupings 55
Decomposer fauna 55
Coprophilous fauna 57
Mycetophagous fauna 57
Necrophilous fauna 58
Outdoor fauna 58
Xylophagous fauna 58
Phytophagous fauna 58
Hydrophilous fauna 58
Synanthropic fauna 58
Medicinal fauna 58
Stored product fauna 58
6.2 Discussion of the Îlot Hunt insect fauna 58
Discussion of the c. 1850 insect fauna 59
16D25 privy fauna 59
Level 16C74 from 16C26 privy section 59
16E100 structure 59
Discussion of the c. 1860 insect fauna 60
16D203 drain 60
16D200 privy section 60
Level 16C70 from 16C26 privy section 61
Discussion of the c. 1870 insect fauna from 16C26 privy section 61
Discussion of the c. 1875 insect fauna from the 16D200 privy section 61
Discussion of the c. 1880-1900 insect fauna 62
16D200 privy section 62
Layers related to the renovation of the 1875 building 62
Layers filling a depression below 1875 building 62
What should we consider an infestation? 63
Discussion of *Lytta* cf. *vesicatoria* 63

Comparison of Îlot Hunt with the Abiel Smith School and Faubourg Québec sites 64
6.3 Interpreting the parasite remains 65
Explaining the absence of *Trichuris* 66
6.4 Nineteenth-century health and sanitation at Îlot Hunt from an environmental, textual and material perspective 67
6.5 What new evidence can this study shed on the nineteenth century? 69
Chapter Seven- Conclusions **71-72**
7.1 Restatement and consideration of hypothesis 71
7.2 Concluding comments on the use of the Îlot Hunt methodology 72
References Cited **73-83**
Acknowledgements **84**
Appendix A **85-122**
Appendix B **123-139**
Appendix C **141-153**

TABLES FOUND THROUGHOUT TEXT

Table 3.1	Summary of levels from 1993 and 1995 excavations	32
Table 3.2	Material culture related to hygienic and sanitary practices recovered from the 1995 Îlot Hunt excavations	33
Table 4.1	Step-by-step methodology for archaeoentomological analysis	38
Table 4.2	Step-by-step explanation of the dilution technique for archaeoparasitological analysis	38
Table 5.1	Archaeoentomological sample sizes and volumes	53
Table 5.2	*Trichuris* counts from the Îlot Hunt archaeoparasitological analyses	54
Table 6.1	Habitat requirements of Micropeplidae, Staphylinidae, Hydrophilidae and Histeridae identified at the Îlot Hunt site.	56
Table 6.2	Ecological groupings of identified insects from Îlot Hunt	57

FIGURES FOUND THROUGHOUT TEXT

Figure 5.1	Most abundant Coleopteran families from Hunt site archaeoentomological investigations	52
Figure 5.2	Introduced, native, and holarctic insect faunas from Îlot Hunt	52
Figure 6.1	Most abundant Coleopteran families from the Abiel Smith School site	64
Figure 6.2	Most abundant Coleopteran families from the Faubourg Québec site	64

APPENDIX A- FIGURES

Figure 1	Location of the Îlot Hunt Site (CeEt-110)	86
Figure 2	Plan of 16D200/16C26 privy and drains, Îlot Hunt Site	87
Figure 3	The Îlot Hunt Site in 1845	88
Figure 4	Sub-operation 11C and plan of 11C11 drain and 16D200 privy section	88
Figure 5	Québec City, in front of Notre Dame Basilica in Upper Town, during the 1832 cholera epidemic	89
Figure 6	Plan taken from 1879 Hopkins Atlas of Québec City indicating the city's wards	90
Figure 7	View of the port of Québec City in 1833	91
Figure 8	Cadastral lots of the Îlot Hunt Site	91
Figure 9	Plan of the Îlot Hunt Site in 1744	92
Figure 10	Plan of Îlot Hunt in 1815 indicating construction of a house on rue Saint-Antoine	92
Figure 11	Southeastern view along rue Saint-Antoine showing houses dating to c. 1815 and c. 1824	93
Figure 12	1824 plan showing second house on rue Saint-Antoine	94
Figure 13	Îlot Hunt in 1824	94
Figure 14	Jean-Baptiste Larue's plan of Îlot Hunt and proposed extension of rue Saint-Antoine in 1829	95
Figure 15	1845 plan of Îlot Hunt	96
Figure 16	Îlot Hunt in 1875	96
Figure 17	Plan of Îlot Hunt Excavations (1988, 1991, 1992, 1993, 1994, 1995)	97
Figure 18	Plan of excavated structures (1988, 1991, 1992, 1993, 1994, 1995)	98
Figure 19	11B50/11B55 covered drain at eastern end of 16C26 privy section	99
Figure 20	Sub-operations 16C and 16D, First Dauphine Battery (1707-1709) and columns from 1875 structure	100
Figure 21	16C26 privy section	101
Figure 22	Location of level 16C74 in 16C26 privy section	101
Figure 23	Eastern edge of the de la Chesnaye wharf (16D101) and First Dauphine Battery (16D100)	102
Figure 24	Profile of 16D25 in southern wall of 16D sub-operation	103

Figure 25	Sub-operation 16D, interior of 16D200 privy section. Stratigraphic profile (facing west) taken form north-south berm	103
Figure 26	16D200 section of privy, southwards view	104
Figure 27	View (facing west) of northeastern corner of 11C15/11C35/16D200 privy section	104
Figure 28	Sub-operation 11C, containing privy structure 11C15/11C35/16D200. Stratigraphic profile from western edge of 1993 excavation.	105
Figure 29	Flown blue plate marked "Daly's Hotel Montréal"	105
Figure 30	Interior of 16D200 privy section with 16D106 column. Stratigraphic profile of southern face of east-west berm	106
Figure 31	Interior of 16D203/16K300 drain	107
Figure 32	Destruction of First Dauphine Battery (16D100) to construct 16D203/16K300 drain	108
Figure 33	Sub-operation 16K, stratigraphic view of west wall	109
Figure 34	Sub-operations 16E and 16F	110
Figure 35	Post-excavation infilling of 16E100	111
Figure 36	Structures related to sanitary practices c. 1850	112
Figure 37	Structures related to sanitary practices c. 1860-1875	113
Figure 38	Transfer print pitcher from level 16D17	114
Figure 39	Brushes and toothbrushes from Îlot Hunt	114
Figure 40	Profile view of brush 16C53-1 dating to c. 1870	115
Figure 41	Medicinal bottle 16C53-2 dating to c. 1870	115
Figure 42	Medicinal bottle 16D17-128 dating to c. 1875	116
Figure 43	Vial 16C70-3 dating to c. 1860	116
Figure 44	Vial 16D35-2 dating to c. 1860	117
Figure 45a	Head of *Anotylus rugosus* (Fabricius)	118
Figure 45b	Pronotum of *Nausibius clavicornis* (Kugelmann)	118
Figure 45c	Head and pronotum of *Sitophilus granarius* (Linné)	118
Figure 45d	Left elytron of *Polygraphus rufipennis* (Kirby)	118
Figure 46a	Pronotum of *Bembidion tetracolum* Say	119
Figure 46b	*Blatella germanica* (Linné)	119
Figure 46c	Pronotum of *Blatella germanica* (Linné)	119
Figure 46d	Head and pronotum of *Cimex* cf. *lectularis* Linné	119
Figure 47a	Left elytron of *Trox scaber*	120
Figure 47b	Head of *Dryophthorus americanus* Bedel	120
Figure 47c	Pronotum of *Creophilus maxillosus* (Linné)	120
Figure 47d	Left elytron of *Necrodes surinamensis* (Fabricius)	120
Figure 48a	Right elytron of *Omosita colon* (Linné)	121
Figure 48b	Head of *Tenebrio molitor* (Linné)	121
Figure 48c	Pronotum of cf. *Lytta vesicatoria*	121
Figure 48d	*Lytta vesicatoria* Linné	121
Figure 49a	Left elytron of *Cercyon unipunctatus* (Linné)	122
Figure 49b	Head of *Ptinus* sp.	122
Figure 49c	Head and pronotum of *Lathridius minutus* Linné group	122
Figure 49d	Right elytron of *Dermestes lardarius* (Linné)	122

APPENDIX B- TABLES

Table 1	Insect remains recovered from privy deposit 16C74 dating to c. 1850	124
Table 2	Coleopteran remains recovered from privy 16D25 dating to 1825-1875	125
Table 3	Coleopteran remains recovered from 16E100 structure dating to c. 1850	125
Table 4	Insect remains recovered from privy section 16D200 dating to c. 1860	125
Table 5	Insect remains recovered from privy deposit 16C70 dating to c. 1860	128
Table 6	Coleopteran remains recovered from 16D203 drain dating to c. 1860	129
Table 7	Insect remains recovered from 16C26 privy section dating to c. 1870	130
Table 8	Coleopteran remains recovered from 16D200 privy section dating to c. 1875	132
Table 9	Late nineteenth century insect remains recovered from privy section 16D200	134
Table 10	Coleopteran remains recovered from late nineteenth century renovations to the 1875 building	135
Table 11	Coleopteran remains recovered from depression fill below 1875 building.	136
Table 12	Insect remains recovered from the Abiel Smith School Site, Boston	137
Table 13	Coleopteran remains recovered from the Faubourg Québec Site, Montréal	138

CHAPTER ONE

INTRODUCTION

1.1 INTRODUCTION TO THE PROJECT

Nineteenth-century urban life in on the eastern seaboard of North America was difficult for many individuals and fatal for some. Epidemics of cholera, smallpox, influenza, and typhoid devastated urban populations. Lack of clean water was one of the most serious problems that contributed to a high mortality rate, particularly amongst the very young. Few cities had adequate if any means of garbage disposal, and usually there were only sporadic health services at the municipal level. The directors of these services often lacked the funds and the necessary political power to make effective improvements to city life.

City populations were growing through immigration and the development and centralisation of manufacturing industries. In cities like Québec and New York, some immigrants settled permanently while others remained only temporarily to earn further passage or to convalesce. Immigrants who were sick or diseased were the responsibility of their host cities which, in turn, had to ensure proper inspection and, if necessary, quarantine restrictions on incoming vessels and passengers. The expanding manufacturing industry in eastern North America further burdened city services as rural populations moved to urban areas in search of factory employment.

The second half of the nineteenth century was also a period of significant advances in medical knowledge, in the promotion of domestic sanitation and personal hygiene, and in the development of urban infrastructures related to health and sanitation. The microscope, developed around mid century, aided the study of organisms that would soon be identified as germs and bacteria, making possible such findings that would revolutionise medical knowledge about disease transmission. The development and promotion of the newly discovered germ theory of 1877 (Cartwright 1977:140), was a critically acclaimed important medical discovery, although by no means was it immediately universally accepted. Germs were now understood to be the cause of diseases that were previously thought to be spontaneously generated, transmitted by noxious fumes or created by miasmic emanations. With germs recognised as disease agents, doctors and other health advocates could prescribe prophylactic measures to hinder the spread of epidemic diseases, while, without this knowledge, they could only treat the resulting symptoms.

Earlier in the nineteenth century there was a growing sanitary reform movement that advocated personal and domestic cleanliness. These domestic sanitarians (Tomes 1990:510) and health reformers were influenced by a similar movement that began in Britain in the early 1830s but had not grown to significant proportions in eastern North America until the latter half of the nineteenth century. Germ theory fuelled the reformers' calls for change in sanitary and medical services. Community groups, largely made up of middle class women and medical practitioners, were instrumental in disseminating and promoting a message of moral superiority through good sanitation and hygiene.

The installation of water and sewage networks in the latter half of the nineteenth century affected the daily lives of most city dwellers, although these networks were not always the improvement which they claimed to be. While some form of infrastructures may have been previously initiated, it was not until the late 1800s in cities like Toronto and Québec that water and sewage services, or some facsimile thereof, were available to most citizens.

Lack of garbage disposal was a further urban blight. Before the installation of sewer networks, most city dwellers used privy pits for their waste disposal. Not only did these pits receive human excreta, they were all-purpose receptacles of domestic waste, which was also thrown into rivers, yards, and vacant lots. Once sewer networks were built, privies were still used for household garbage, but the pits themselves were often connected to the sewers via drains, which often clogged. Eventually municipal politicians would demand the pits be put out of use. Previous to this, many cities employed people to cart waste out of towns for use as fertiliser, but an official municipal disposal service with strict regulation was badly needed. While garbage services did not always parallel the development of the sewer and water systems, an unfortunate situation for many people, this period generally witnessed the genesis of some form of garbage disposal.

To implement and oversee these new services, in co-operation with other city sectors, municipal health departments were established. While boards of health were generally formed to deal with the crises of epidemic diseases throughout the nineteenth century, permanent Boards of Health with salaried medical officers were not formed until later in the century. Their jurisdiction generally covered the creation and enforcement of sanitary by-laws, quarantines, milk and meat purity, and public inspections during epidemics. These offices were also consulted about the installation of new city infrastructures such as sewer and water networks and garbage disposal.

Nineteenth-century Québec City fits the model outlined above. Health reform and major public works such as sewage and a municipal water service were begun during this period. They were partially instigated by the cholera and typhus epidemics that ravaged Québec throughout the 1900s (Bilson 1980:114). Despite sanitary advances made in the city and changing medical and sanitary practices promoted by health reformers, doctors and the Municipal Health Board, Provincial Hygiene Inspectors' reports from the late nineteenth century cast the city in a different light. Québec City was portrayed as a filthy place with garbage disposal unchecked, abundant and overflowing privy pits, contaminated water and insufficient sanitary services. The Lower Town area of the city was further singled out due to its appalling conditions (Dagneau et al. 1983:162). While the urban infrastructure advances, medical and sanitary practices, and sanitary regulations should have created a cleaner Québec City, the Provincial Hygiene Inspectors' reports indicated otherwise.

Québec City was always geographically, economically and culturally divided into Upper and Lower Towns (Appendix A, Figure 1). The Upper Town traditionally contained the city's religious and government institutions and the homes of the elite. Lower Town, wedged between the base of the steep Cap-aux-Diamants cliff and the St. Lawrence River, was the economic hub of the city, focussed primarily around the port, with its many wharves and depots. In the nineteenth century, the port area was the centre of trade, banking and the shipbuilding industry, employing a great number of people. The banking centre of the city was located on rue Saint-Pierre, and other businesses in the area included professional trades and artisans. Many middle-class people, such as bank managers and merchants worked in the Lower Town district but lived in Upper Town. The inhabitants of Lower Town were business owners, artisans, and labourers living near their places of employment. When the port industries began to decline in the 1860s (Hare et al. 1987:262) and the manufacturing sector expanded, many labourers left the port area, to move closer to their new jobs in the Saint-Roch ward.

The Lower Town of the city, houses the Îlot Hunt site (CeEt-110), the focus of this study (Appendix A, Figure 1). The site is representative of Québec City's history as it contains components of military, naval, commercial, industrial, and domestic. The land for this site was initially ceded to Charles Aubert de la Chesnaye in 1687 (L'Anglais 1998:ix) and the site has been continuously occupied to the present, though it is the period between 1850-1900 that is the scope of this project, a time of significant sanitary advances in the city.

The site was intensively occupied during the nineteenth century, a period which coincided with the economic prosperity of the City. This period is well represented at Îlot Hunt by the construction of both domestic and commercial structures, a stable, a possible well, and two privies, one of which is part of a larger sanitary system with connecting drains. The large privy and an associated drain (Appendix A, Figure 2), excavated in 1995, and dating to the latter half of the nineteenth century, that make up the majority of this sanitary system, has provided much of the biological and artifactual material under consideration.

This project will consider sanitary conditions and the health and medical practices of the Îlot Hunt site inhabitants during the latter half of the nineteenth century. Material cultural remains, in the form of structural remnants and artifacts along with textual or documentary information are used to reconstruct the sanitary state of the site. Archaeologically recovered insect remains and parasite eggs are examined to gauge whether they confirm or contradict the understanding of the site developed by the more traditional areas of archaeological knowledge: texts, artifacts and architecture.

To assess this information, a multidisciplinary approach that combines textual, archaeological, and biological data is used. The textual or documentary record informs us on the broader subjects of sanitation, medical knowledge and practice and is used to reconstruct the demographic profile of the site and site history. The material culture also reflects the medical beliefs, sanitary practices and economy of the site. The biological data involves the examination and interpretation of preserved insect remains and parasite eggs. As archaeoentomology and archaeoparasitology are relatively new approaches in North American archaeology, they shall be discussed in detail in this chapter, as will the documentary and material culture sources used. This interdisciplinary approach, examining the many relevant sources of information or contexts of the Îlot Hunt site between 1850-1900, will aid in interpreting the sanitary reality experienced by the inhabitants of the Hunt site in its urban historical context and will provide information about the health and medical practices of its users.

1.2 AN INTERDISCIPLINARY ANALYTICAL APPROACH TO STUDYING ÎLOT HUNT

In this project, it is essential to integrate the site into its historical and environmental settings using the tools of historical archaeology and the techniques of environmental archaeology. A multi-disciplinary approach is used in studying the Îlot Hunt site, analysing artifacts, texts, structures, and biological data, which are accessed through archaeoentomology and archaeoparasitology.

An ideal approach for the Îlot Hunt site study is modelled after that defined by Beaudry and Mrozowski in their documentary, environmental and archaeological analysis of a nineteenth-century New England mill town in Lowell, Massachusetts (Beaudry 1989; 1993; Beaudry and Mrozowski 1989; Mrozowski et al. 1989). At Lowell, a multi-disciplinary approach was developed to study the use of domestic space, health and hygiene, foodways, consumer patterns, and corporate control amongst the mill workers living in company housing at the Boott Cotton Mills (Beaudry and Mrozowski 1989:2-4).

According to Beaudry, the approach used in the Boott Cotton Mills project is contextual using Hodder's definition of context as well as Butzer's notion of the dynamic ecological context (Butzer 1980, 1982). Beaudry describes this approach as an integrated, interdisciplinary archaeology with an interpretive analytical approach (Beaudry 1989:1, 1993:91). This rightly considers and discerns the many contexts to be considered, with recognition of belief systems, cultural, historical and environmental contexts (Beaudry 1993:91). The interpretive approach uses the constructive elements from a variety of viewpoints and "eschews narrow or rigid theoretical positions that tend toward ecological functionalism, economic determinism, or doctrinaire application" (Beaudry 1993:91).

In his discussion of environmental contexts Butzer states that, too often, the environment is treated like a constant variable, and is in this way, taken for granted (Butzer 1982:4; see also O'Connor 1998:4). Furthermore, environmental information should not be seen as descriptive background data, but as a dynamic factor in the analysis of archaeological contexts. Analysis of context is one of the primary themes for several approaches in archaeology, due to input from cultural anthropology, human geography, and biological ecology (Butzer 1982:12). Butzer is critical of other approaches to environmental archaeology such as Evans' approach (see Evans 1978). He states that Evans simply describes techniques, and their application in environmental archaeology which he considered too technically focussed rather than centring on the goal of the research at hand (Butzer 1982:5). According to Butzer, the ultimate goal of research is to always determine the interrelationship between culture and the environment (Butzer 1982:5). Butzer generally uses this active and dynamic view of the environmental context in analysing prehistoric sites. While in this view environmental context is given considerable treatment, an approach that properly integrates textual and other data is required.

Hodder identifies the need to draw on extensive, detailed documentary research involving as broad a range of sources as possible to aid in defining *any* object's context in archaeological studies. Hodder states that objects can be mute when out of their 'texts', but that most archaeological objects are, by definition, situated in time, place and relation to other objects, essentially their network (Hodder 1987:2). We, as archaeologists, are charged with reading this network or context of the objects. It is this notion of context or the "totality of relevant dimensions", which must be read or analysed (Hodder 1987:5).

The Îlot Hunt site has many contexts that are at once urban, historical, environmental, cultural and economic, and may be accessed through studying relevant artifacts, texts and biological data. To balance the interpretive process in the analysis of archaeological data, the role of the environment or environmental context should not be singularly considered or given any preferential priority in analyses as the data provided by documents and material culture data is equally as important. According to Mrozowski, archaeologists "can ill afford to ignore any avenue of inquiry, be it biological or social analysis" (Mrozowski 1988:22). The treatment of all data therefore must be interpretive to properly place this information in its rightful place respecting the other contexts without being skewed by a rigid theoretical framework.

To consider the multiple lines of evidence to be used in studying Îlot Hunt, an approach modelled after that of the Boott Cotton Mills project (Beaudry 1989; 1993; Beaudry and Mrozowski 1989; Mrozowski et al. 1989), contextual, interpretive, and interdisciplinary, is proposed. An approach of combined documentary, artifactual, and environmental evidence has been demonstrated by other researchers to be an effective means of studying historic sites of northeastern North America (Fries et al. 1990; Geismar 1993; Horne and Tuck 1996; Howson 1993; Reinhard 1994; Reinhard et al. 1986). The Îlot Hunt project has many sources of information relevant to this study that must be considered and this analysis seeks to use a number of constructive approaches. I hope that by studying the historical, material, and environmental contexts of the site in detail, an integrated, approach can be applied which will provide new insights into nineteenth century life in Québec City. As Deagan points out, environmental studies in historical archaeology concerning health and sanitation have been successful in providing new insights into urban life, due to the integrated, multifaceted, environmental approach they employ (Deagan 1996:370).

1.3 ARCHAEOENTOMOLOGY- CURRENT TRENDS

Archaeoentomology is the study of insects from archaeological deposits in order to reconstruct the cultural and natural environments of sites at the time of deposition (Elias 1994:107). Certain characteristic body segments of microsculpture are used for identification, whilst the genitalia, if preserved, provide absolute species identification. However, if these entomological remains are poorly preserved and lack the characteristics needed for identification or if the taxonomy or ecology is poorly understood, interpretations can be difficult. Beetles (Order: Coleoptera) are the most commonly used insects for archaeoentomological analyses, although bugs (Hemiptera), flies (Diptera), bees, ants, wasps (Hymenoptera), moths (Lepidoptera), fleas (Siphonaptera), lice (Anoplura), and mites (Acarina) are also recovered. Most of the focus of archaeoentomology has been on Coleoptera, since they are normally better preserved than other insect orders due to their highly sclerotised exoskeletons. Furthermore, they can be identified to species using taxonomic keys and modern collections in conjunction with microscopic characteristics on the head pronotum/thorax and elytra/wings (Coope 1970:99; Morgan and Morgan 1980b:1106).

As Coleoptera are one of the most populous groups on the planet, they have been extensively collected and studied for at least the last two centuries in North America, and even longer in Europe. Their known ecological requirements and taxonomic relationships make them one of the most sensitive biological indicators (Osborne 1988:715). Other groups such as flies (Skidmore 1997) and mites (Schelvis 1987, 1997) have been studied in archaeological settings, though not as widely as beetle-based research. Regional and local environments can be accurately reconstructed, as many beetle species are stenotypic (i.e. restricted to a particular habitat), resulting in an adaptation to a precisely defined environment. It is this characteristic above all others that makes Coleoptera such valuable palaeoecological indicators (Lowe and Walker 1984:182). Also, the ecological requirements of many beetle species have not shifted over time (Coope 1970:107; Kenward 1976:8; Morgan and Morgan 1980b:1106) allowing for the interpretation of past environments and insect distributions by comparison with modern ones. If we consider insects representative of their environmental contexts, then, by extension, we may consider insects representative of their archaeological contexts, where they may indicate specific ecologies as well as cultural activities.

Furthermore, unlike mammals, there has been no evidence of significant extinction amongst Coleoptera for at least the last two million years, solidifying their value as environmental proxy data (Elias 1994:59). In the past, insect remains observed in geological and archaeological soil samples were largely ignored, as it was believed they were continually evolving and were therefore not analogous to the present fauna (Morgan and Morgan 1987:38). Conversely, beetles have not undergone significant speciation or evolution in Britain and North America during the Quaternary period (about the last 2 million years) (Coope 1970:101, 1978:185; Kenward 1976:8; Matthews 1981:1098), and therefore modern insects used for identification are recognised as analogous to the fossil specimens. Due to the combined factors of evolution, ecology and entomological knowledge, we can use preserved beetle remains to evaluate archaeological sites with the confidence that they can be properly identified to the species level.

While palaeoentomology studies insects from past environments, the archaeoentomological approach, studying preserved insects from archaeological sites, was first used in British archaeology in the 1960s and 1970s (Buckland et al. 1974; Coope and Osborne 1968; Kenward 1975a, 1975b, 1976, Osborne 1969; 1971). Extensive archaeological research has been undertaken at York, England where, due to the rising water table, over five successive metres of organic material from the Roman to the Late Medieval Period have been exceptionally well-preserved (Buckland et al. 1974:25). The research, experimentation and publications based on York and other sites in Britain served to establish the archaeoentomological method, although in North America this research is in its pioneering phase (Elias 1994:191),as its potential as a useful technique is slowly gaining recognition (Buckland et al. 1995; Dirrigl and Greenberg 1995; Schwert 1996). North American studies have been primarily conducted in the field of palaeoentomology, where preserved insects from geological contexts are used to reconstruct glacial movements and subsequent floral and faunal repopulation of the landscape (Matthews 1981; Morgan and Morgan 1980a, 1980b, 1987; Schwert and Ashworth 1988).

Archaeoentomological studies in North America have been, until quite recently, concerned with prehistoric sites in the central and southwestern United States. Most of this research has been on adjacent 'natural' environments rather than on anthropogenic deposits (Elias 1994:124). Thus far, studies have indicated the role of insects in the prehistoric diet, the seasonality of timber preparation by Anasazi Indians and burial preparations in Colorado and South Dakota (see Elias 1994:125-129). Archaeoentomological research on historic period sites has been undertaken only in northeastern North America and includes the Cross Street Back Lot (Bain 1998b; Bain and LeSage 1998) and Abiel Smith School Sites (Bain 1998a) in Boston, Massachusetts, Îlot Hunt (Bain 1997a), the Faubourg Québec site in Montréal (Bain 1997b), and the Ferryland Site in Newfoundland (Bain 1997c).

Environmental reconstructions

Beetle-based studies on glacial environments have enjoyed a successful thirty years of research in Britain where the approach was conceived, while similar research in Eurasia, North America and South America only began during the last twenty years. Much of the work has been devoted to the study of the beetle fauna related to the four glacial interglacial cycles of the Quaternary period (Atkinson et al. 1987; Coope 1996; Coope and Brophy 1976; Elias 1994; Schwert and Ashworth 1988; Schwert et al. 1985; Morgan and Morgan 1980b). Holocene environmental reconstructions, some of which document human impact on the landscape, are becoming increasingly more common (Bain 1998b; Elias 1995; Lavoie et al. 1997; Morgan 1987; Osborne 1988; Schwert 1996). These studies have improved our understanding of both climatic change and the environmental impact of human settlement in North America and Europe.

A landmark study in glacial palaeoentomology was Russell Coope's (1973) analysis of English palaeoentomological material from the last glaciation. Coope recovered over 100 individuals of a dung beetle, *Aphodius holdereri* which presently inhabits the high Tibetan plateau at altitudes of 3-5,000 m asl, and therefore lives in a cold thermal environment. Similar beetles have been recovered from sites throughout the English Midlands that date to the last glaciations (42,000 B.P. and 26,000 B.P.). The presence of *A. holdereri* during these mid-Devensian (Wisconsinan) deposits indicates the temperature regime of that time period and the changing nature of insect distributions since the Pleistocene (Coope 1973:335-6). Previously, the effect of glacial episodes on beetle fauna had been only speculative. Coope's study on beetle fauna demonstrated the dynamic nature of faunal change as a result of glacial episodes.

In Morgan and Morgan's work (1987) on the fauna of the Sangamonian interglacial on Baffin Island, which began around 130,000 years ago and lasted for 11,000 years (Roberts 1989:42), the beetle species revealed a mean July temperature 5-6° C higher than the present (Morgan and Morgan 1987:40). Without these results, typical temperatures estimated for this period were thought to be two degrees *lower* than present, due to the lack of human-induced change seen primarily in rising carbon dioxide levels (Morgan and Morgan 1987:40). This study indicates how beetle species may reflect slight temperature gradients, and can refine our understanding of past environments.

Similar types of studies have dominated palaeoentomological research and indicate climatic shifts as brief as a thousand years (Morgan and Morgan 1987:38). In general, the palaeoentomological record has demonstrated that beetles respond quickly to climatic shifts, due to their ability to colonise new landscapes rapidly seeking out suitable environmental niches (Coope 1970:112; Coope and Brophy 1976:125; Jones 1986:27; Morgan and Morgan 1987:40), highlighting their value in environmental reconstructions.

In the investigation of late a Bronze Age riverine deposit,

Osborne (1988) examined samples dating from around 3000-2500 B.P. The presence of coprophagous dung beetles and beetle larvae that live on grass roots indicated adjacent grassland grazed by livestock such as sheep or cows (Osborne 1988:716-721). Climatically, the site appears to have been slightly warmer than today, as indicated by the abundance of *Onthophagus* dung beetles which are uncommon today in southern England, but are increasingly more abundant in Europe. As the climate cooled slightly, *Onthophagus* became less abundant and beetles belonging to the genus *Aphodius* increased. These are commonly encountered dung beetles at present. By studying the relative abundance of these two genera from several sites and combining this data with the known habitats and present distributions of these beetles, Osborne was able to demonstrate that the site's environment was slightly warmer than today (Osborne 1988:725). Interestingly, these findings contradict results of one of his earlier studies (Osborne 1969:562).

Schwert, in his study of late Holocene insect faunas from Northeastern Iowa (Schwert 1996), reconstructed local environments while documenting change in the insect fauna since European settlement in the mid-western United States. In his analysis of organic sediment from six locales in the Roberts Creek drainage basin, Schwert was able to detail the drastic change in the fauna throughout the historic period, while the three pre-settlement faunas that span 2900-150 B.P. showed a certain faunal constancy (Schwert 1996:55).

Insect biogeography

Another focus of research in archaeoentomology has been the study of insect introductions or biogeography (Bain 1998b; Bain and LeSage 1998; Buckland 1976, 1981, 1988; Buckland et al. 1995; Kislev 1991; Osborne 1971; Pals and Hakbijl 1992; Panagiotakopulu and Buckland 1991; Sadler 1991; Sadler and Skidmore 1995; Schwert 1996). The term 'introduction' in this context refers to a process whereby a species is introduced to a new environment through both intentional and unintentional human action. Many of these introduced species are classified as storage pests of agricultural products. In Britain, this fauna was almost entirely introduced and dispersed during the Roman period via their complex grain shipping network (Buckland 1982:233-234). In North America, the introduced beetle fauna is primarily Eurasian in origin and is believed to number around three hundred species (Buckland et al. 1995:227). The shipping trade during the historic period created suitable environments for many beetle species, allowing them to survive the transatlantic crossing (Sadler 1991:200). The use of European farming and pasture systems in North America led to an essentially Neo-European landscape (Crosby 1986:167), with introduced insect species moving into the newly available ecological niches in North America where they had very little competition (Buckland et al. 1995:237).

Reconstruction of agricultural practices and storage estimates

The ability of beetles to respond to changing environmental conditions is aptly demonstrated by the Coleopteran fauna found in the excavation of a Roman well in Belgium (Ervynck et al. 1994). In this study, 16 species of carabids or ground beetles from five levels detail Roman occupation, intensive agricultural activity and the subsequent field abandonment. The oldest layer shows pasture grassland while the following two layers indicate an agricultural landscape that was perhaps farmed with only a single crop (Ervynck et al. 1994:263). Finally, the top two layers document abandonment and reversion of the land to a fallow field. The authors feel that these radical faunal changes document the rapid depletion of agricultural soils during the Roman occupation of Gaul (Ervynck et al. 1994:264).

In palaeo-economic reconstructions, grain production and storage estimates have been used to reconstruct past populations and economies (Halstead 1990; Jones et al. 1986). However, Buckland (1978:43) has shown this to be a flawed approach, unless the inevitable presence of storage pests is considered as part of the ancient agricultural economy. Furthermore, Buckland has also suggested that the expanded cultivation of the Roman period was not attributable to the *Pax Romana*, nor to military needs, but instead to the increased presence of aggressive pests such as the granary weevil. This and other pests of stored grain could have accounted for the loss of at least ten percent of all grains produced (Buckland 1978:45).

Health and sanitation

The archaeoentomological record also allows us to study medical practices and sanitary conditions of past sites and their inhabitants via certain species of beetles, bugs, fleas and lice. For example, documentary and archaeological evidence has indicated that the beetle species *Lytta vesicatoria* or Spanish Fly was used in medical treatments ranging from premenstrual syndrome to injured fingers and feet to sore throats (Anonymous 1895; Bain 1997a; Hakbijl 1986, 1987; Hakbijl and de Groot 1997; Moore 1874).

Sanitary indicators such as the human flea (*Pulex irritans*), head lice (*Pediculus humanus*), pubic lice (*Pthirus pubis*), and bedbugs (*Cimex lectularius*), all considered ectoparasites, have been recovered during archaeological investigations (Bain 1997a; Buckland 1987; Buckland and Sadler 1989; Gill and Owsley 1985; Girling 1984; Kenward 1999; Panagiotakopulu and Buckland 1999; Schelvis 1994; Sveinbjarnardóttir and Buckland 1983). While the presence of a louse in an archaeological assemblage does not specifically imply that the inhabitants were diseased or very dirty, lousy people are, more at risk to infections from bites (Busvine 1976:51-55). A similar tenet holds for the presence of insects that attack stored products, a relatively common find in archaeoentomological studies (Bain 1998b; Buckland 1981; Kenward and Williams 1979; Osborne 1981; Panagiotakopulu and Buckland 1991). The occasional presence of such individuals may be expected, but at times they have been found to completely dominate an assemblage, indicating potentially serious health and economic implications for those fed by such stores (Kenward and Williams 1979:73). Health problems related to the presence of stored product pests may include skin reactions, allergies, intestinal dyspepsia, and the transmission of pathogens (Robinson 1996:237). Toxic microflora caused by the beetles' metabolic activity may also render food inedible. A remarkable example of pest infestation was found in studying levels from the Roman occupation of York, England (Hall and Kenward 1976). Samples from what was determined to be a grain store contained approximately 100,000 individual insects per kilogram of deposit (Hall and Kenward 1976:276). As no grains themselves were preserved, this overwhelming evidence of an infestation would not have been recognised unless a multidisciplinary approach was used in the study of the environmental remains. Archaeoentomological analyses of storage pests, bedbugs, fleas, and lice augments our ability to understand past sanitary conditions, although Osborne (1983:461) cautions about assumptions placed on interpretations of such infestations if we impose twentieth century Western standards of hygiene on our interpretations of the past.

Answering archaeological questions

The preceding discussion has demonstrated that preserved entomological remains from archaeological sites may be used to interpret the environment, and economy of a site itself and the health of its inhabitants. Specific questions often arise in archaeological research and may be asked of archaeoentomological assemblages. Such questions may concern the function of specific structures, the presence of unusual layers, the validity of generally-accepted hypotheses, and the nature of the faunas themselves. The following two examples demonstrate how such archaeological questions may be answered.

Buckland (1982), tested an historical assumption concerning possible evidence, in the form of burnt layers, of destruction and conflagrations due to barbarian movements into northern England in A.D. 296. At the Roman fort of Malton in East Yorkshire, a large burnt grain layer was presumed to indicate this intended burning and evacuation (Buckland 1982:54). However, following the examination of the charred material, Buckland discovered that there was an infestation of the grain by three of the principal grain storage pests (Buckland 1982:56). As a result, he suggested a potentially sounder interpretation for this and many of the burnt layers at Roman sites throughout Britain and elsewhere: deliberate burning of storage residues to rid storehouses of infestations by stored product pests (Buckland 1982:56). While Buckland cannot prove definitively that his alternative hypothesis is correct, by simply examining the charred material microscopically, he was able to challenge an untested archaeological assumption.

The historic City of York, England, with its exceptional organic preservation and a research mandate promoting interdisciplinary research, has been the site of many studies using insect remains (Buckland 1976a, 1976b; Buckland et al. 1974; Hall and Kenward 1976, 1990; Hall et al. 1980, 1983; Kenward 1978a; Kenward and Large 1998; Kenward and Williams 1979). As an enormous volume of material has been studied, many problem-oriented questions have been applied to the data. To address the difficulty of interpreting insect samples, the York researchers have developed methods of comparison by ecological groupings, abundance rankings, diversity indices, and statistical treatments of faunas. Over twenty-five years of intensive research has focussed on understanding the complexity of both the urban natural and urban cultural environments in the Roman and Anglo-Scandinavian periods. They have addressed such issues as the visibility of the indoor versus the outdoor component of insect faunas, the concept of the house fauna, insect death assemblages and seasonality of occupation.

Archaeoentomology is a powerful approach to interpreting archaeological sites that has gone, until recently, almost unnoticed in North American historical archaeology. The economy, environment, and health of a site's occupants may all be accessed via the investigation of the preserved beetle fauna in contexts which range from the regional setting of the archaeological site to the contents of a single layer within a structure such as a privy or stable. For this reason this approach has been chosen in the investigation of the Îlot Hunt site in Québec City with the hopes of better understanding the nineteenth-century health, sanitary and medical practices of the site's inhabitants.

1.4 ARCHAEOPARASITOLOGY- CURRENT TRENDS

Archaeoparasitology reconstructs human sanitary practices and the natural and cultural environments of archaeological sites via the analyses of preserved parasite eggs or larvae. In the study of ancient diseases, many illnesses leave no trace on human bones (Jones 1982:66); sampling and testing for preserved parasite eggs may give previously unknown visibility to these afflictions. Parasite studies may also indicate the presence of animals and their relationship to human occupants, socio-economic conditions, demographic information about the site's inhabitants, agricultural practices, and food storage, preparation, service and disposal as well as health, disease and site formation processes (Driscoll 1994:8-10).

Archaeoparasitology generally used helminths classified as tapeworms (cestodes), roundworms (nematodes), thorny-headed worms (acanthocephalans), and flukes (trematodes). These all have durable eggs and are the most likely to be recovered archaeologically (Reinhard 1992:233). The parasites we are interested in studying are helminths or worms, which are parasites of vertebrate animals (Chandler 1955:239).

Many parasites need very specific conditions to live, while others are free-living. Host-parasite relationship may be complex with one or two intermediate hosts (Reinhard 1992:233-4). Some parasites are able to exploit a number of hosts, a situation that occurs with whipworms, where human and pigs can be cross infected with the same species (Beer 1976:47). Other species are highly adapted and connected to the host in such a way they will survive when its host is in any environment, even in extreme temperatures (Reinhard 1992:234).

The two most commonly encountered parasites from archaeological contexts in Europe and North America are the nematodes *Ascaris lumbricoides*, also known as the intestinal roundworm and *Trichuris trichiura* or whipworm (Greig 1982:51). While in small numbers these parasites are benign, they can be considered indicators of faecal-borne diseases (Geismar 1993:63), and their presence may provide insight into sanitary practices (Reinhard 1992:239). *T. trichiura* adults grow about 3-5 cm in length and live in the region of the cecum, at the junction of the small and large intestine, and in the appendix (Chandler 1955:394). Severe infestations may result in anaemia, fever, nausea, diarrhoea, loss of appetite, and emaciation (Chandler 1955:395). Mature *A. lumbricoides*, can grow to 15-31 cm, and live in the small intestine, though in the immature stages, they pass through the heart, liver, lungs, trachea, throat, oesophagus and small intestines (Chandler 1955:442-3). In great numbers they can cause general abdominal discomfort, vomiting, diarrhoea, a mild fever, and in extreme cases they can block the entire small intestine, or cause appendicitis and pneumonia (Chandler 1955:444-445). Other parasite species are found only occasionally in archaeological settings, likely due to their morphology. For example, *Trichinella spiralis*, a nematode species that exists in inadequately cooked pork, has never been found in archaeological deposits, as its eggs are not resistant to decay (Jones 1982:68).

There are several important factors to consider when assessing parasite eggs from historical contexts. The first is that interpretations that can be derived from archaeoparasitological data are dependent on the demographic knowledge of the site including the socio-economic status, occupation and size of population (Reinhard 1994:63). Secondly, soil samples must come from faecal deposits, which can be difficult to identify archaeologically, however, an abundance of small seeds such as *Rubus* or raspberry are indicative of faecal-laden soils (Fortin 1996:4,11). By determining the abundance of seeds correlated to dietary pollen, it can be discerned whether a level is mostly faeces or trash (Reinhard 1992:240), a useful factor in both parasite and archaeological studies. The amount of egg decomposition must also be assessed. Eggs are durable, but can decompose in some situations (Reinhard 1994:63). Studying the relative condition of pollens, seeds, and other remains aids this assessment. Finally, the faecal deposits must be in a datable context. For more accurate interpretations, short time frames are preferred (Reinhard 1994:63).

Parasite research in archaeology was recognised as a potential source of information in the 1960s, but was not fully developed as an archaeological investigative approach in Europe, South America and United States until the 1980s (Reinhard 1990:162). North and South American researchers have studied both indigenous and European sites dating to the historic period, though the historic period studies are primarily located in northeastern North America. Pre-Columbian studies in the United States (Faulkner 1991; Hall 1977; Horne 1985; Reinhard 1988, 1990; Reinhard et al. 1987) and South America (Allison et al. 1974; Horne 1985) involved mummies and preserved faecal and intestinal remains, commonly known as coprolites. Research in the Northeast has been entirely based on privy contents (Driscoll 1994, 1995; Fries et al. 1990; Geismar 1993; Horne and Tuck 1996; Howson 1993; Mrozowski et al. 1989; Reinhard 1994; Reinhard et al. 1986).

Literature review of archaeoparasitological studies

Bog bodies, mummies, coprolites, and privies have all been the subjects of parasitological investigations. As many species of parasites have been recovered from these diverse contexts, time periods, and geographical settings, selected research will be briefly discussed to indicate the breadth of data derived from archaeoparasitological studies. Privy studies from historic sites in northeastern North America will be treated in the following section, as they address specific questions regarding sanitation in the historic period of the Northeast.

In peat preserved bog bodies such as Lindow Man found in Cheshire, England, whipworm and roundworm have been identified, while other bog bodies from Denmark, Poland, and Germany revealed similar gut parasite remains (Brothwell 1986:73-76). Desiccated remains of hydatid cysts, or tapeworms, were found in England, Scotland, and Denmark (Jones 1982:66). Hydatid cyst disease is one of the most debilitating diseases caused by helminths (Reinhard 1992:239).

A study of the faecal remains of a Greenland Inuit mummified woman yielded evidence of pinworm (*Enterobius vermicularis*) infestation (Lorentzen and Rørdam 1989:143), generally an indication of crowded and unsanitary living conditions (Reinhard 1992:238). A potential case of false parasitism was noted in a mummy recovered from the Alaskan permafrost, where evidence of a fish worm egg (*Cryptocotlye lingua*) was found in faecal remains and was probably preserved in the digestive track (Horne 1985:304; Reinhard 1992:236). South American mummies have yielded evidence of both the *Trichuris trichiura* and *Ascaris lumbricoides* (Allison et al. 1974:104; Horne 1985:301-3).

Archaeological studies of human coprolites from North and South America, Poland, Israel, Germany, and Austria have also revealed the same groups of parasites found in bog bodies and mummies. In the Americas, prehistoric coprolite studies or palaeoscatology, has been carried out on both human (Faulkner et al. 1989; Hall 1977; Horne 1985; Moore et al. 1974; Reinhard 1988; Reinhard et al. 1987) and animal faeces (Araújo et al. 1982; Horne 1985). Mexico, Peru and the American Southwest have been the source of much of the coprolite research. Horne (1985:306) points out that the population of the New World was infected with all three phyla of worms that cause disease, though studying them is somewhat restricted as arid climatic conditions are needed for their preservation.

Cave coprolites from Tennessee, Kentucky, and Peru contained eggs of the roundworm *Ascaris lumbricoides* (Faulkner 1991:687; Horne 1985:301-2). The Tennessee coprolites also contained evidence of the human pinworm *Enterobius vermicularis*, and the first New World evidence of *Giardia* (Faulkner 1991:687). *E. vermicularis* was also found in Oregon coprolites (Hall 1977:9). Humans are the only hosts of pinworms, and the archaeoparasitological record indicated that the populations of the prehistoric Southwest were frequently infested (Horne 1985:301). Due to their reproduction, with eggs being laid in the perianal region, evidence of them is found in only five per cent of faecal remains of infested persons (Horne 1985:301), so their frequent presence in coprolites in the Southwest indicates a heavy parasite burden.

Flukes have been identified from coprolites in Chile. One fluke species, *Paragonimus caliensis* has a complex life cycle including crayfish or shrimp that act as a second intermediate host (Horne 1985:304). Evidence of hookworm infestation (*Ancylostoma duodenale*) has been found in a Peruvian mummy (Allison et al. 1974:104). Evidence of such parasitic infestations in the Americas indicate their presence pre-dated European contact, an important biogeographic consideration in the history of human parasite interactions. Coprolites from Chile, Peru and Utah also contained traces of tapeworm infection (Horne 1985:305), while thorny headed worms, or Acanthocephala, have been identified from Utah and Oregon. These may cause severe damage and even death, as they perforate the gut wall by attaching themselves, or cause peritonitis, a disease in which the lining of the abdominal cavity becomes inflamed (Hall 1977:10; Horne 1985:305).

An interesting coprolitic study undertaken by Reinhard tested the hypothesis that the adaptation to agriculture in the southwestern United States has led to higher rates of helminth infection and parasite diversity (Reinhard 1988:356). He suggested that factors such as sedentism, the use of faecal remains as fertiliser, proximity to a damper environment, and switching to a diet lacking anti-helminthic properties would have led to higher levels of parasites in the local population. Parasitic evidence from sites before and after the adaptation to farming supported this hypothesis (Reinhard 1988:356).

Privy deposits from sites dating from the Roman to Medieval periods have been investigated from northwestern Europe and Britain. They generally indicate infestation by the commonly encountered roundworms and whipworms as well as liver flukes, tapeworms and horse pinworms (Jones 1982:67-8). Refuse on archaeological sites with some degree of faecal content are extremely common, though they also contain a wide range of other types of refuse (Jones 1985:110-1).

The examples presented here indicate the rich diversity of parasites that appear in coprolites, privies, and bodies in the Americas, Europe, and Israel. Many parasites have complex stages of reproduction, which must be properly understood to assess their presence and significance in the archaeological record. Archaeologists in northeastern North America have studied parasites eggs preserved primarily in privies, and their work has contributed to understanding attitudes about and the changing practices of health and sanitation in the last two centuries.

Archaeoparasitology in Northeastern North America

Archaeological models that combine documentary evidence supplemented with archaeological and parasitological data have been used to interpret past health and sanitation on several historic sites in north eastern North America. Geographically this area covers Newfoundland (Horne and Tuck 1996) and the states of New Hampshire (Driscoll 1994), New York (Geismar 1993; Howson 1993), Massachusetts (Driscoll 1995; Mrozowski et al. 1989), Rhode Island (Reinhard et al. 1986), Maryland (Driscoll 1994), Delaware (Fries et al. 1990) and West Virginia (Reinhard 1994). These studies have yielded information that may be classified as parasitological analysis at the household level, while making general statements about hygiene and sanitation, chronic parasitism of site inhabitants, and resistance to improved sanitation. These archaeoparasitological studies will provide a comparative framework for the results of the Îlot Hunt site analysis.

Parasitological analysis at the household level

Two eighteenth century privies from Queen Ann's Square in Newport, Rhode Island (Reinhard et al. 1986), indicated discrepancies between parasitic infestations at two households, while elucidating information about other domestic practices such as gardening.

The Brown household contained a militia captain and merchant, while the Tate household housed a blacksmith and his wife, a seamstress (Reinhard et al. 1986:31). There was evidence of roundworm, hookworm, tapeworm, and high levels of whipworm infestation in the Brown household privy, while the Tate household contained some whipworm and high levels of roundworm (Reinhard et al. 1986:33-34). Therefore, both homes were parasitised but in different ways. Pollen and seed evidence suggests that the Brown home consumed more commercially-grown plants, while the Tate household home-cultivated plants for consumption (Reinhard et al. 1986:34). Both the roundworm and whipworm are transmitted through contact with faecally-contaminated materials, with a common route of infection being poorly washed vegetables grown in contaminated soils. *A. lumbricoides*, the roundworm has a sticky uterine coat that adheres to growing stems of plants, and is easily transmitted by the consumption of plants grown in open garden plots which are exposed to sunlight. Therefore the higher amounts seen in the Tate privy indicated their use of a domestic garden (Reinhard et al. 1986:34).

Resistance to improved sanitation

At Harpers Ferry, West Virginia, soils from three areas were examined specifically for the presence of *Ascaris lumbricoides* and *Trichuris trichiura*. Privy use was from the nineteenth into the twentieth century (Reinhard 1994:63). Although, there was poor organic preservation in the twentieth century privy, there was consistent evidence for unsanitary conditions generally, and of faecal-borne disease into the twentieth century (Reinhard 1994:65), a time when parasite levels should have dropped due to better sanitary conditions and infrastructure improvements. These consistent parasite levels demonstrate, according to Reinhard, resistance to improved sanitation, was also seen at a Wilmington, Delaware site (Fries et al. 1990).

Evidence for chronic parasitism

At Wilmington, four labourers' privies, dating from 1790-1920, were analysed (Fries et al. 1990:18). Roundworm and whipworm were the predominant parasites in all privies, particularly those privies that pre-dated the construction of the city's sewer system in 1890. Two other parasites, *Fasciola hepatica* and *Dicrocoelium dendriticum* may have been accidental human infections or livestock contaminants (Fries et al. 1990:23-4). The constant parasite presence indicates the chronic infection experienced by the site's inhabitants (Fries et al. 1990:25).

At the Cross Street Back Lot site of Boston, a large Colonial privy was used by multiple family generations. In her analysis of the parasite remains, Driscoll found evidence of consistent worm infection throughout fourteen of the fifteen layers analysed (Driscoll 1995:4), and concluded this was another potential case of chronic infestation.

Other valuable information can be derived from this microanalysis of archaeological levels. In her analysis of the Cross Street privy, Driscoll noted that levels which were originally interpreted as fill levels, actually contained the remains of preserved parasite eggs (Driscoll 1995:4). If the fill layer can be identified as originating from the general yard area, perhaps its re-examination may provide more information about outdoor urban conditions.

As a tool that can access information about sanitation, human behaviour, hygiene, and health, archaeoparasitology is a powerful approach. Preserved parasite eggs or ova have been studied from a variety of sources including historic privy contexts not unlike the Îlot Hunt site privy. While a number of studies have been undertaken in northeastern North America, this will be the first use of this approach in Québec.

1.5 STUDYING HEALTH AND SANITATION

To reconstruct the sanitary setting of the site and of its occupants from 1850-1900, documentary sources provide information about many of the site's contexts. They detail the history of the city, the Îlot Hunt site, its occupants, and their medical and sanitary situation. The material culture recovered provides further information about medical and sanitary beliefs and practices. Material culture also includes site structures and architecture relevant to this investigation of the Îlot Hunt sanitary system.

The documentary record

The city of Québec has been the subject of many detailed studies that portray the economic, social, militaristic, and religious life of the city since its founding in 1608. The main sources consulted for this study of the city were *Histoire de la ville du Québec* (Hare et al. 1987), *A Tenant's Town* (Desloges 1991), and *Québec, 1765-1832* (Ruddel 1991). As many studies have been undertaken at Îlot Hunt (Bouchard 1997, 1998; Boucher 1999; Cloutier 1991; deVarennes 1994, Dubé 1995; Goyette 1994, 1999; Groupe Harcart 1989; L'Anglais 1993, 1998; Leclerc 1992, 1995, 1998; Moussette and Auger 1987; Simoneau 1987; Simoneau and Rouleau 1988), primary documents were consulted for years and subjects that had not previously been treated.

Works which addressed the medical history of Québec were, *La médecine au Québec: Naissance et Évolution d'une profession* (Bernier 1989), and Goulet and Paradis', *Trois siècles d'histoire médicale au Québec: chronologie des institutions et des pratiques* (Goulet and Paradis 1992). The sanitary history of Québec City is treated in *La Ville de Québec, histoire municipal* 17, 19, 35 (Dagneau et al. 1983; Drolet 1965; 1967) and Pothier's *L'eau, l'hygiène publique et les infrastructures* (Pothier 1996). A nineteenth century sanitary and medical history of Québec amongst other Canadian cities was provided by Bilson's *A Darkened House: Cholera in Nineteenth Century Canada* (Bilson 1980), Tétreault '*s Les maladies de la misère, aspects de la santé publique à Montréal 1880-1914* (Tétreault 1983), and Brace and MacDougall's articles (Brace 1995; MacDougall 1982) on the sanitary history and infrastructure of Toronto. Duffy's *The Sanitarians: A History of Public Health* (Duffy 1990) discusses the history of sanitary institutions and reforms in the United States.

Historical accounts of nineteenth century medical practices were taken from Duffin's *Langstaff: A Nineteenth-Century Medical Life* (Duffin 1993), and four medical guides and recipe books dating between 1864 and 1895 (Anonymous 1882, 1895; Chase 1864; Moore 1874).

Published historical sources from Québec City detail the municipal by-laws and services available during the nineteenth century. The City of Québec had a health office intermittently during this period and documents dating to 1849, 1854, and 1891-1900 were used to assess the history of city sanitation, which included complaints and legislation about privies and cesspools. Municipal by-laws published by the Québec Fire Society in 1833 and in the *Québec Gazette* for the years 1885 and 1891 were consulted. A valuable late-century perspective on sanitation in the city was gleaned from the Provincial Hygiene Inspectors' reports for the years 1888, 1889 and 1891.

Demographic data providing an economic and social profile of the inhabitants of the study area was drawn from a diverse number of sources. Census of Canada records were consulted for 1831-1901, as were annual city directories for 1858-1871, 1877, and 1880. Assessment roles were also studied for the years 1842-1843, 1847-1849, 1855-1856, 1861, 1863, 1870-1871, 1875, and 1880-1881.

Material culture studies

Material culture in the form of artifacts are an important resource that allows us to reconstruct and better understand the lives of the inhabitants while providing culturally sensitive information (Mrozowski et al. 1989:299). Urban privies are one of the most important features on historic sites as they provide this wealth of artifactual data. The privy system at Îlot Hunt was no exception and yielded a rich and diverse assemblage of ceramics, clothing, hardware and personal items. Artifacts uniquely related to sanitary practices which will be considered are toothbrushes, toothbrush holders, soap holders, shaving materials, water pitchers or jugs, chamber pots, combs, feeding bottles, medicinal bottles or containers, and syringes (Baugher-Perlin 1982:272-273; Howson 1993:148-155; Recherches Arkhis 1987:147-149). Comparative data from contemporaneous sites are also available (Beaudry and Mrozowski 1989; Garrow 1996; Howson 1993; Recherches Arkhis 1987).

The Îlot Hunt sanitary system

The material culture of the Îlot Hunt site consists not just of the artifacts found in the site excavations, but also of the structures of the sanitary system itself. This complex, which provided the artifacts and biological data for this study, is composed primarily of two privy structures 16D25 and 16D200/16C26, and their associated drains. The buildings that likely used these structures include the Hunt House, situated at 66 rue Saint-Pierre, the three houses on rue Saint-Antoine, and the other buildings on the property, which included a stable and warehouses (Appendix A, Figure 3).

Both privies are situated in the 1745 fill layer (16D3) which dates to the French regime and lies east of the de la Chesnaye wharf (16D101) and the First Dauphine Battery (16D100). These two structures are united into one ong stone feature (Appendix A, Figure 1). The 16D25 privy was originally discovered during the 1992 excavations and was labelled 10A10 (L'Anglais 1998:88). The north wall of this structure was exposed during 1995 excavations with a mechanical excavator, being used to clear away fill to reach the ancient beach level in front the Battery (Bouchard 1998:64). The privy is situated in front of the de la Chesnaye wharf which functions as its west wall, while the east wall is constructed of wooden planks. The privy was sampled for biological remains and noted at this time, but it was not subjected to any other systematic treatment or measurement.

The 16D200 section of the 16D200/16C26 privy is composed of a wooden rectangular structure that measures 3.15 m north south and 2.45 m east west. Annexed onto the southeast corner of 16D200 is the extension 16C26 which measures 3.4 m northwest to southeast and is 1.75 m wide (Appendix A, Figure 2). The layers within this two-part structure total 1.1 m thick and represent five separate depositional events (Bouchard 1998:4). Both sections are constructed of wooden planks.

There are two drains connected to the 16D200/16C26 privy. One of these, 16D203/16K300 is connected to the privy on the northwest side of the 16D200 section while the other is attached to the southeast corner of 16C26. 16D203 is 5.25 m long running in an east west direction, measuring 0.7 m wide. The 11B50/11B55 drain is attached to the southeast corner of the 16C26 part of the privy and was excavated in 1993 (Goyette 1994:57-58). A third drain was excavated in 1993 (11C11/11F4) (Appendix A, Figure 4) that may have connected to the 16D200/16C26 privy system. Goyette believes it was used as a drain by the occupants of the middle of the three dwellings on rue Saint Antoine, as it crosses the yard in a southeasterly direction, but has a western arm that almost connects with the east wall of the 16D200 privy. It was constructed around the same time as the privy and was in use until 1860 (Goyette 1999:141). While samples and artifacts were not studied from this drain, its structural presence is of interest in the yard as it is part of the mid-century sanitary system.

1.6 STATEMENT OF HYPOTHESIS

The latter of half of the nineteenth century involved many changes in the daily lives of city dwellers. For many North Americans, connections to water and sewer systems became widely available for the first time and cities instituted health offices and inspectors. This time period also saw the rise of the domestic sanitary or hygienic movement, which promoted cleanliness of the home and person as a means of preventing epidemic diseases such as smallpox and cholera. The combination of water and sewer connections, the promotion of personal, domestic, and municipal sanitation, and improved medical practices should have improved life in urban centres over this fifty year period, however, the image of the nineteenth-century city in northeastern North America provided by documentary evidence does not confirm this. Similar discrepancies between documentary records and material culture were found in the study of Manhattan's Five Points neighbourhood in New York City (Yamin 1997, 1998).

If we consider the changes in medical practices, public awareness and in urban infrastructures during the last half of the nineteenth century, the Îlot Hunt site should have become a progressively cleaner as well as a healthier place to live. Four factors indicate this: the installation of municipal and domestic sanitary structures such as sewer and water networks and indoor plumbing, strict sanitary regulations, changing medical practices, and the influence of health reformers, made up of medical practitioners as well as community groups. The late nineteenth-century drop in mortality rates appears to confirm this. At the Îlot Hunt site, socio-economic analysis indicate the site's economy improved after 1870 (Bouchard 1998:94), and therefore the quality of life of the inhabitants should have paralleled this situation over the fifty-year period from 1850-1900. Textual data presents, however, a conflicting picture of the city and particularly of Lower Town. Even with the advances listed above, Lower Town in late century was still unclean and considered filthy by the Provincial Hygiene Inspectors in their late century reports (Archives nationale du Québec, Québec [ANQ], Rapports d'inspection, Conseil d'hygiène de la province de Québec [CHPQ], 1888, 1889, 1891, 1893, 1898, E88 1971-07-003/21 vol 1 [1887-1891]). But how can we accurately examine whether urban environments were actually physically cleaner, with less refuse to contaminate water and food sources, and whether people were using and practising more sanitary care? As Deagan points out "one of the most important considerations in historical archaeological research is to avoid investigations that simply affirm, repeat or refute information already known through documentary sources. Doing so requires identifying questions that can be informed only through the integration of environmental, documentary, contextual sources with material culture" (Deagan 1996:36). The Îlot Hunt sites has generated such questions, which are addressed in this multidisciplinary project.

My primary hypothesis contends that if we consider what I will call the *promotional perspective* provided by the textual evidence of the municipal and provincial by-laws, the medical community and the urban infrastructures, an overall improvement in late nineteenth century life was experienced, which is confirmed by a reduced mortality rate. If this was the case, then we should see several indices of this on the Îlot Hunt site. These include a progressively reduced degree of parasitism throughout the second half of the nineteenth century, and privy pits covered over and indoor plumbing installed or at least the on-site sanitary structures connected to the city's sewer and water networks. There should also be progressively less insect infestation in houses and foodstuffs, and evidence of fewer generic medicinal remedies represented by medical bottles as more scientifically based medical practices are diffused, and a generally cleaner yard space at the site as people are taking measures to ensure a more salubrious environment.

The secondary or null hypothesis states that considering the *reported perspective*, new hygienic standards are not being integrated into peoples lives and existing sanitary systems are not improving. If this is the case, the degree of parasitism in the population will remain relatively constant throughout the latter half of the nineteenth-century, and there will be the continued application of spurious medical remedies. The yard will indicate trash and refuse strewn about, there will be significant indoor infesting insect faunas, including pests and there will be ongoing evidence for privy use as people did not upgrade their sanitary facilities.

To test these hypotheses, I will study the relevant textual and material data, which includes site structures and artifacts. These results will then be compared to the results of the archaeoentomological and archaeoparasitological analyses. This multidisciplinary approach will provide a more holistic view of the site and its urban setting, and a better interpretation of changing medical and sanitary practices of the nineteenth century.

CHAPTER TWO

URBAN LIFE IN NINETEENTH-CENTURY NORTH AMERICA

Many cities in nineteenth-century eastern North America experienced unprecedented growth and industrial development. This situation demanded the implementation of sanitary policies and structures ranging from quarantine regulations during epidemics, to the tenure of animals within cities, to the construction of complex sewer and water systems. The installation of sanitary systems by municipal governments was a major political issue of this time period (Brace 1995:33; Duffy 1990:77; MacDougall 1982:1), as cholera and other epidemics were a continuing menace to the urban population. These new political initiatives reflected the changing medical and hygienic beliefs of the period. This chapter will discuss life in the urban centres of the Northeast during the latter half of the nineteenth century and the concurrent medical and sanitary developments, beliefs and practices. It will also focus on developments in Québec City from 1850-1900, in an attempt to understand the sanitary and medical reality faced by the inhabitants of the city and, more specifically, of the Îlot Hunt site.

2.1 LIFE IN THE NINETEENTH-CENTURY CITY

The early nineteenth century growth of industry in many cities led to an influx of people who were former rural-dwellers or recent immigrants. With rapidly growing tenements, and without the necessary infrastructure, urban slums were firmly established in New York by the 1830s (Duffy 1990:77). Life in the nineteenth-century city was unpleasant (see Cronon 1991:10), and even fatal, for many people, especially the very young. Epidemics spread rapidly through crowded metropolitan areas and between cities, by train and water traffic (Leavitt 1982:22). While some epidemics, such as typhoid and cholera, were transmitted through water, contaminated water was also the source of other problems such as gastro-enteritis and diarrhoea, one of the primary causes of nineteenth-century infant deaths (Bernier 1989:132; Tétreault 1983:518).

Many North American cities experienced epidemics of smallpox, influenza, yellow fever, cholera and typhus between the 1830s and the 1870s (Duffy 1990:105; Goulet and Paradis 1992:215-220). During these times of crisis, Toronto and Québec City would instate medical officers to conduct inspections and implement strategies for the treatment of the ill and for sanitising the perceived problem areas of the city (Bilson 1980:6). These were curative rather then preventative strategies (MacDougall 1982:2), and were typical of most northeastern North American cites, as these metropolitan areas lacked the permanent infrastructures needed to fully address the problems.

By late century, rapid urban growth and an expanding industrial sector nullified the progress of most sanitary reforms such as water and sewer networks and waste collection. City officials found they could not provide services for all inhabitants (Duffy 199:175; Leavitt 1982:11). Toronto more than doubled in population in the 1880s to 181,220 (MacDougall 1982:4 & 7), and Milwaukee grew from 20,000 to 300,000 between 1850-1900 (Leavitt 1982:3). While the tremendous growth was welcome economically, it overburdened the inadequate municipal services (Leavitt 1982:12). Municipal health departments grew, with many cities expanding or building their sewer and water systems, implementing garbage collection, street cleaning, boards of health, inspectors, and sanitary regulations (Brace 1995:33; Duffy 1990:77 & 147; Leavitt 1982:7; MacDougall 1982:1). While these expanded services may not have greatly improved life in the urban sector, the mortality rates generally declined for the period 1880-1900. An exception was amongst the very young whose mortality rates rose due to the poor living and working conditions experienced by many during the industrial revolution, and by the slow changes in medical practices (Bernier 1989:148).

The poor quality of life in cities led to the initiation of municipal and private campaigns aimed at improving urban conditions, personal cleanliness, and the immoral behaviour of city dwellers (Duffy 1990:81), three themes that were inextricably linked in the nineteenth-century consciousness of middle-class health advocates and practitioners. It is believed that the late-century drop in the urban mortality rate seen in some regions of the United States and Canada was due to the rise of the hygienic or sanitary reform movement of the 1870s (Brace 1995:34; Duffy 1990:147). Their proposed reforms reflected the 'sanitary idea' that was conceived in England in the 1830s and 1840s as a reaction to the Industrial Revolution's urban social problems (Duffy 1990:66). It was an approach using investigation, legislation and administration, which demanded a motivated and actively involved government (Brace 1995:35), something that was initially lacking in North American cities. Many of the first changes advocated by such groups were, as a result of the lack of government structure, health and diet reforms, and were aimed mostly at individuals instead of communities (Duffy 1990:66).

Urban populations needed systems to dispose of human wastes and domestic garbage. Privy pits were the most common form of human waste disposal in cities and towns, and became one of the most difficult of all sanitary problems to overcome. Municipal by-laws generally specified how and where they should be constructed, as well as dictating the frequency of their cleaning and inspections (Duffy 1990:73; Goulet and Paradis 1992:234; Leavitt 1982:4). When water and sewer connections became available later in the century, closing them over and discontinuing their use was supposed to be the normal procedure, though this was not always the case. For other garbage, most cities had designated areas although rivers, streams, streets, vacant plots of land and backyards appeared to all have become receptacles in most cities. Duffy notes that public institutions such as hospitals often let their wastes flow into rivers or drains (Duffy 1990:88).

By 1830 most large American towns regulated the construction and location of privies (Duffy 1990:73), although in cities such as Pittsburgh attempts at sanitary legislation were "cheerfully disregarded" (Duffy 1990:69). Although the first privy by-law in Québec City was created in 1673 (Saucier 1958:46), over two hundred years later it appears people were still generally ignoring such by-laws, which still were not enforced. In 1829, New York City legislated that privies had to be five feet deep and made of stone or brick and could not be built within thirty feet of a public well (Duffy 1990:73). The size, depth, building materials, as well as the distance from water sources, along with stated intervals for their cleaning, were typical guidelines for privy construction.

New York City took action to regulate privies early in the century. Theirs was a particularly dire situation with up to twenty families sharing a single pit (Duffy 1990:73). In 1802, a municipal workforce was to be created that would disinfect the city's privies with lime, and the following year, an office of the Superintendent of Scavengers was created to control and schedule privy cleanings (Geismar 1993:60). Also in 1803, 14 inspectors, two from each of the city's seven wards, were appointed to oversee that the privies were being emptied (Geismar 1993:60).

Poorer landlords were not likely to spend money on privy maintenance for their tenants, so there were frequent overflows (Duffy 1990:88). Even when linkages to sewer and water services were available, people did not readily embrace these changes and the ensuing costs, partially due to the medical beliefs of the time. At the Greenwich Mews site in New York City, Geismar observes that the site's privies were likely built around 1844, and were in use at least until 1880, though sewer service was available in 1853 (Geismar 1993:62).

In Toronto in 1870, most homes still had a privy pit in their yard, connected to a drain which often discharged into a river or stream or they emptied them manually (Brace 1995:34). With the increasing promotion of the new 'sanitary ideas' from Britain, people petitioned for sewer construction, a proposal often supported by the City Engineer (Brace 1995:37). In 1885, after inspecting two-thirds of the city's homes, health workers found that 60 percent of households still used privy pits. As it was costly to remove them completely, the city's Medical Health Officer recommended they be cleaned and disinfected annually (Brace 1995:38). Other cities like Boston and New York hired scavengers or local farmers to remove the privy's contents after cleaning (Duffy 1990:88). Initially, this material was carted out of cities for use as fertiliser, although at times, the waste was simply abandoned on vacant lots (Duffy 1990:73). In Lowell, Massachusetts this practice was phased out due to the nuisance of leaky carts (Mrozowski et al. 1989:306).

A Montréal study indicates a direct correlation between the presence of privy pits and an elevated mortality rate. In the Sainte-Marie ward, 87 percent of families still used privy pits in 1887, and the mortality rate was 35.7/1000, whereas in Saint-Antoine, only 35 percent of families had privies and the rate was 18.6/1000, the lowest rate in the city (Tétreault 1983:524). The provincial mortality rate at this time was 21.9/1000 (Bernier 1989:115). Tétreault states that the Montréal Health Office used all possible methods to abolish the use of the privy pits, and it was this abolition that must have accounted for the lower mortality rates in some wards, although, in certain sectors of the city, he also attributes this to the insufficient sewer system (Tétreault 1983:524).

Lowell, Massachusetts, a corporate mill town, seems to have been an exception as garbage was strictly regulated and regularly removed, by the mill corporation. Ironically though, contaminated water was still drunk from the canal which also had privies emptying into it (Mrozowski et al. 1989:307). This example demonstrates a lack of knowledge of disease transmission, and the inconsistent application of sanitary practices of the nineteenth century.

In many cities, access to sewer and water networks came in stages, with wealthier neighbourhoods receiving the first access to, or being the first recipients of these services. In Toronto, the 1859 procedure to have a sewer installed involved two thirds of a street's residents signing a petition. The value of their properties had to add up to at least half the assessed value of the properties affected (Brace 1995:34). As the city did not expand the sewer network systematically, neighbourhoods with the economic means could effectively petition for sewer installation.(Brace 1995:34).

Sewer and water systems were not always constructed as two parallel systems as they should have been, and therefore did not always function properly. In Montréal, the drains built in the first half of the nineteenth century to remove rainwater had only slight or weak slopes, and were not well sealed (Trépanier 1996:68). When citizens decided, often independently, to use them as sewers for domestic wastes, this resulted in faecal build-up and soil saturation when they backed up into cellars. This worsened as the populations grew and water-closets were installed, adding more volume to the system (Trépanier 1996:69). This water system began as a private venture in 1800, but was under municipal control by 1845. It did not provide sufficient or constant water and was at times supplemented by water carriers (Drolet 1967:30). In 1852, 90 percent of Montréalers did not have running water in their homes (Pothier 1996:40), though by 1897, only seven percent of the population continued to share one tap between three or four families (Pothier 1996:42)

Water quality was a problem in most cities. The contaminated water was sometimes used to water down milk and, as a result, caused infant diarrhoea and the spread of cholera and typhoid (Tétreault 1983:522). This practice led to the regulation of the milk supply. Few city dwellers had the means or space to tenure cows, often banned in cities due to the waste they generated. As a result, people had no alternative but to purchase milk.

The newly-created nineteenth-century boards of health were established to control problems like the tainted milk supply. Their activities involved inspecting and quarantining houses, public health education about hygiene and epidemic diseases, quality control for milk, meat, and ice, and the development and expansion of the garbage, water and sewer systems (Leavitt 1982:4; MacDougall 1982:6).

New York City and the state of Massachusetts established boards of health in 1866 and 1869 respectively. These represented a departure from the normal laissez-faire approach to urban sanitary problems, as they achieved full departmental status within the urban and state governments (MacDougall 1982:3). A more typical approach was taken by the cities of Montréal and Trois-Rivières, when faced with the smallpox epidemic of 1885 (Tétreault 1983:511). To combat this problem, they erected temporary smallpox hospitals that were closed after the epidemic ended (Goulet and Paradis 1992:248-9), not an uncommon approach in treating epidemic diseases in the 1800s. Montréal's Board of Health had an uncertain start. It was officially established in 1877, was abolished in 1879, and then re-established in 1881. In the interim, the city suffered through epidemics of typhoid and scarlet fever (Goulet and Paradis 1992:243) without the guidance of a central health authority. Toronto appointed its first permanent salaried medical officer in 1883. This office soon became very active, using the British technique of house-to-house inspections to initiate public education of the health office, and to create a list of sanitary offences, which would then raise the politicians' awareness (MacDougall 1982:5). Previous to this, medical officers had been primarily political appointments, reflecting the current political will and popular medical beliefs of constituents (Duffy 1990: 85; MacDougall 1982:2).

Not all cities had the fiscal means to address their sanitary needs. Québec City underwent a recession in the 1870s, unlike Montréal and Toronto, which both experienced growth during this period. During the 1880s, Toronto devoted, on average, $30,000 annually to health care (Dagneau et al. 1983:155). The average amount allocated by Montréal at this time was $20,000 (Dagneau et al. 1983:155), for a population which would surpass a quarter of a million over the next decade (Bernier 1989:149). Québec City spared only $1,000 (Dagneau et al. 1983:155), a paltry sum even though the city's population was about a quarter of Montréal's. A stumbling block faced by many cities was the lack of commitment to invest in the necessary infrastructures of sewers, water systems, sanitation services, and paving. Politicians were more predisposed to pay for short term care of the sick and the poor, than the costly, long term investment of infrastructure works (MacDougall 1982:3). Many city dwellers were left to make their sanitary reforms individually as, at times, there was only a primitive public hygiene service (Duffy 1990:123; Tomes 1990:510), and written legislation did not always ensure implementation.

Nineteenth-century cities faced many challenges, and the establishment of urban infrastructures such as garbage, sewer, and water networks did not necessarily mean the use of these systems by all citizens. Furthermore, popular notions of illness during this period hindered the progress of sanitary reforms, although by the late nineteenth century the urban mortality rate was dropping, new discoveries in medicine were gaining wider acceptance, and more than ever, urban populations in northeastern North America had access to sanitary services.

2.2 NINETEENTH-CENTURY MEDICAL DEVELOPMENTS, BELIEFS AND PRACTICES

In the early nineteenth century, a doctor's main function was the alleviation of the patient's symptoms. Due to the state of medical knowledge of the period, illnesses themselves were rarely treated (Bernier 1989:117). The body was viewed as a single entity, made up of easily identifiable structures that were related in an orderly fashion. Diseases, therefore, were treated as they affected these structures (Cartwright 1977:135), instead of attempting to understand the root cause of the disease. Through medical discoveries later in the century, germs and bacteria were isolated as the causes of disease and infection, although this was a controversial development which was not universally accepted.

Nineteenth-century medical and sanitary practices were dominated by a few causation theories of disease. The most popular ones were based on beliefs about the atmosphere and environment, morality, and noxious 'miasms' or gas-like odours (Bernier 1989:130). Most of these ideas had their roots in the eighteenth century, though ideas about miasms had existed since the classical period. With the development of bacteriology and germ theory later in the nineteenth century, more scientific views were proposed.

Eighteenth-century atmospheric explanations for disease proposed that air, water, heat, cold, humidity and the electric state all had an influence on diseases (Bernier 1989:130; Duffy 1990:67). Physicians would keep meteorological records during epidemics looking for these associations. The concept of 'zymotic' diseases, those caused by impure air and water likely stemmed from these early beliefs about the atmosphere. Cholera, typhoid, diphtheria, smallpox, measles and scarlet fever fell into this classification (Tomes 1990:518). The contagionist view declared that 'contagions' were intrusive living microscopic particles that could multiply in their host and in air and water (Goulet and Keel 1993:419). The proper environmental conditions would cause contagions to become active and to lead to outbreaks of contagious diseases (MacDougall 1982:2), thus supplying an explanation for the spread of illnesses. When cholera struck Québec City, barrels of tar were burned in the evenings to remove the disease 'contagion' that was thought to be in the air (Bilson 1980:34).

Morally-based theories drew on the observations that diseases often struck the poor neighbourhoods with more severity than wealthier areas. The lifestyle (drinking, gambling on Sundays and Mondays) of the less fortunate and the working classes was therefore questioned (Bernier 1989:130). Boards of health emphasised cleanliness and moderation in alcohol combined with a healthy diet to combat illness. As so many of the early victims of the nineteenth century cholera epidemics were poor, it was easy to take this moral position of them (Bilson 1980:33). So great was the perceived relationship between drinking, moral weakness, and disease, when cholera struck Québec in 1832, the taverns were immediately shut (Bilson 1980:37).

The most popular and enduring causation theory was that of the centuries-old 'miasmic' explanation. This long-held belief stated that decaying organic material such as human and animal wastes or putrefying vegetable matter emitted noxious vapours or 'miasms' which caused disease (Cartwright 1977:20; MacDougall 1982:2). How this miasma originated and spread was not a topic which doctors agreed upon (Tomes 1990:518). The miasmic thesis was proposed as an explanation for what appeared to be spontaneous epidemics such as cholera and yellow fever, diseases that were not transmitted by contact (Duffy 1990:68). When swampy tracts of land were drained and settled in the United States, it was believed that the ensuing malaria and other illnesses were due to the miasma that was emanating from the ground once they were exposed to the sun's rays, poisoning the atmosphere (Duffy 1990:67-68). A physician in early nineteenth-century Wilmington, Delaware suggested that gaseous wastes of the nearby marshes, and slime or putrefaction from rotting vegetables caused the dysentery he witnessed that summer (Fries et al. 1990:25).

Regulations published by the Central Health Office of Québec in 1854 reflected these commonly-held beliefs. To prevent another epidemic of cholera, the immediate cleaning of all yards, cellars, and manure was to be undertaken. If waste could not be removed, sand was to be used to dry up such messes. Houses were to be cleaned and aired out, and cellars that functioned as kitchens should also be aired out day and night. Walls were to be whitewashed twice a month during the epidemic, and the floors of houses without carpets should be washed and brushed twice a week. During the epidemic, people were to sleep in the highest rooms of their home for better air circulation. Any rotting animal or vegetable remains were to be disposed of immediately, and in a manner that assured public health. Finally, any source of stagnant water near houses, yards, streets and in vacant lots was to be immediately drained or dried up with sand (Bureau central de santé 1854:3, 9). Even after the cause of many of the epidemic diseases was discovered later in the century, foul odours from excrement and rotting domestic wastes were believed to be harmful, so pervasive was this miasmic thesis. While not scientifically proven, these three main concepts of disease propagation all pointed to improved urban sanitation and sanitary reform as a preventative cure (Shryock 1966:173).

By mid-century, Virchow recognised that the body is made up of cells which are individuals derived from one another, and that diseases cause them to distort at an abnormal speed, a new discovery which affected the current perception of disease (Cartwright 1977:135). Building on this discovery, Pasteur and Koch's work in 1876 and 1877 demonstrated that germs are a cause of disease and not a product of it (Cartwright 1977:140). Koch, by 1881, had hypothesised certain relationships between bacteria and disease, the result of which stated that a disease is the result of a specific organism (Cartwright 1977:141). Once identified, it was possible to discover how these organisms were transmitted and the role the carriers played in spreading the disease. These new discoveries about bacteria formed the basis for the sanitary reform movement, while research on germs, disease causation, and disease transmission, formed the tenets of what is referred to as germ theory (Duffy 1990:193).

When the new ideas of germ theory were beginning to be disseminated, the foundation was already laid for its acceptance by the domestic sanitary reform movement that promoted a sense of personal responsibility in disease prevention (Tomes 1990:528). For many individuals, it provided the scientific basis for the reforms they had always been promoting (Tomes 1990:529). The concept of the germ was assimilated easily into older notions of decay and putrefaction, as the 'contingent-contagionist' view that stated under certain conditions, disease germs in rotting privy or sewer contents could be activated which would lead to diseases (MacDougall 1983:2). This view conveniently embraced both ideas while advocating for cleaner cities and homes.

These new notions of germs continued to fuel the cause of the domestic sanitarians (Tomes 1990:510), who claimed that damp cellars, poor ventilation, dirty carpets, filth, and unclean living caused the spread of typhoid, scarlet fever, diphtheria and other infectious diseases. These beliefs were rapidly disseminated through popular literature, magazines, newspapers, and advice books (Tomes 1990:510). As municipal sanitary structures were often primitive, these print resources were aimed at private homeowners and encouraged people to initiate their own personal preventative measures (Tomes 1990:510). There were four main areas of domestic conduct for the sanitarians: the proper construction and maintenance of the house, the nursing of the infected, the hygiene of the nursery, and general housekeeping (Tomes 1990:523). In the United States, this movement had its roots in the early 1800s and normally involved middle and upper class women, and the medical profession (Duffy 1990:68). While bacteriology and germ theory revolutionised medicine and public health, the domestic sanitary movement made a major contribution to community medicine (Duffy 1990:193). It was based on British reforms, which led to the 1848 Public Health Act of Britain and the establishment of a national central board of health (MacDougall 1982:2). The British public health movement stressed environmental sanitation which was clean water, effective waste management and good housing (MacDougall 1982:2), as well as, good general health through nutrition and immunisation (Brace 1995:34; Shryock 1966:27).

Even with the discovery of germs, the miasmic thesis continued to underlie many beliefs of the sanitary reform movement and played a role in public health until the early twentieth century (Duffy 1990:129). The public clung to this notion in ways that today seem rather bizarre. For example, in the early 1900s some American municipalities never undertook street excavations during the summer months as they were thought to expose the population to disease (Duffy 1990:129). In 1885, a text within the Report of the Montréal Natural History Society stated that "there is nothing that pollutes the air so quickly and with such injurious effects as human *excreta*. Nature has wisely and mercifully made all such matter offensive to the sense...Wherever there is an offensive odour there is danger". The article goes on to state that this poisoned air kills many citizens in Montréal each year (Report of a meeting of the Montréal Natural History Society 1885:7-8). A company, who manufactured dry indoor latrines or pan-closets, purported to be odourless, and therefore not a health risk, published this treatise. Such publications would have fuelled the debate between germ theory and other more traditional views.

In 1870s Toronto, sanitary reform advocates were initially doctors and medical practitioners, but the notion quickly spread into acceptance in the middle class (Brace 1995:37). The "mouvement hygiéniste" or sanitary reform movement began in Québec around 1825 and was primarily focussed on the plight of the poor as they had the highest mortality rates (Bernier 1989:117), a cause that remained the movement's emphasis into the late nineteenth century (Guérard 1996:208). They were made up of the province's medical elite who organised themselves as a political unit to better the lives of the working class, employ preventative measures against contagious illnesses, and educate the public, the same reforms being promoted in many industrialised countries (Deschênes 1981:356, 374). The Montréal Sanitary Association was founded in 1867, followed by the Citizen's Public Health Association in 1875 (Goulet and Paradis 1992:235, 240). One of North America's greatest advocates was Québec's Dr. Severin Lachapelle who became the president of the American Public Health Association in 1893 (Goulet and Paradis 1992:259). He published the *Manuel d'hygiène à l'usage des écoles et des familles* in 1888, and the *Manuel d'hygiène à l'usage des écoles et des familles rédigé conformément aux instructions du Conseil d'hygiène de la Province de Québec* in 1890 (Goulet and Paradis 1992:253-256). The English translation of Lachapelle's 1890 treatise was published in 1891, and in it we find statements like 'dirtiness always breeds disease' (Lachapelle 1891:159), and 'eternal vigilance is the price of everything worth the having or keeping' (Lachapelle 1891:169). Such heavy-handed messages were created to inspire guilt and anxiety (Tomes 1990:527).

Québec's universities, Université Laval and McGill University, were abreast of the new bacteriological developments, but institutional responses were slow, and the practice of the bacteriology was not seen until the turn of the century, (Goulet and Keel 1993:419). At McGill, a diploma in public health and sanitary science was created in 1899. By the year 1908, hygiene courses were offered for engineers, architects and military medical candidates (Goulet and Paradis 1992:261).

The smallpox epidemic of 1885 focussed the energies of Québec's hygienist movement, and resulted in the publication *Journal d'hygiène populaire*. The medical profession and the sanitary associations joined forces to pressure the Provincial Sanitary Office to help with implementing sanitary legislation (Dagneau et al. 1983:160). By 1880-1900, the sanitary revolution was in full swing in many North American cities. As a result, water and sewer systems were established, regular garbage collection and removal was implemented, and health departments expanded (Duffy 1990:175).

Even once the mechanisms and sanitary treatments were better understood for the major epidemics, there were still prejudices to overcome. At the turn of the century hospitals were still thought of as places to go to die, and few people wanted to send sick family members there (MacDougall 1982:5). While different medical theories and sanitary practices were applied, and municipal services varied from city to city, there was an overall drop in infectious diseases in the 1890s due to these changes (Brace 1995:38). However, conflicting beliefs during this crucial development period of public sanitation hindered sanitary progress, and life in cities continued to be a struggle for many.

2.3 LIFE IN QUÉBEC CITY 1850-1900

Québec City was typical of many North American cities facing the nineteenth-century hurdles of instituting sanitary reforms, allocating necessary funds, and encouraging a compliant and responsible population. During its rapid expansion in the first half of the nineteenth century, Québec City was faced with repeated waves of epidemic diseases, a lack of garbage disposal, and poor drinking water, issues that were regulated but appear not to have been enforced. After 1850, the city created a permanent Board of Health, constructed water and sewer networks, regulated privy construction, waste disposal, and quarantines, as well as general sanitation requirements for living in the city. This covered topics such as snow removal, yard clean up and the tenuring of animals within city limits. While such topics were regulated, the public's action regarding such measures is harder to gauge and the late century hygiene inspectors' reports do not present a favourable image of the city.

The first of many cholera epidemics struck eastern North America in 1832, and was followed by repeated waves until 1873 (Bilson 1980:3). Cholera was not the only epidemic illness of this period as the inhabitants also endured smallpox, influenza and typhoid (Goulet and Paradis 1992:202).

The image of Québec City in 1826 and 1832 (Appendix A, Figure 5), as it was preparing for the first cholera epidemic, is bleak. Garbage was dumped on riverbanks within the city where it became landfill (Hare et al. 1987:211). There were many open sewers, no regular street-cleaning, and after the spring thaw, debris from the previous winter was simply left in the street (Drolet 1965:68). Both open and closed sewers made of wood and ceramic, drained into the Saint-Charles and St. Lawrence Rivers. This was not, however, unique to Québec City as open sewers and drains were commonplace in many North American cities (Duffy 1990:76). Water sellers took their water from the polluted riverbanks and sold it door to door, although families with wells were not immune from contamination either, as water from privies leached into wells and cisterns (Hare et al. 1987:211). Around this time, the author Susanna Moodie described Québec City as a filthy hole, which looked better from the ship than it did on shore(cited in Ruddel 1991:8).

The city was therefore ill prepared when the repeated epidemics of cholera began in 1832. It struck again in 1833, 1834, 1839, 1849, 1851, 1852, 1854, 1866, 1871, and 1873. Cholera claimed 3000 victims in 1832, 1282 in 1833, and another 2000 in 1834 (Drolet 1967:71; Hare et al.1987:210). The severity of the outbreaks varied, but 1849 and 1854 were particularly bad (Bilson 1980:114; Dagneau et al. 1983:153). The first Canadian health office was created in Québec City in February 1832 (Drolet 1965:67), in order to prepare and enforce quarantine legislation. It was primarily made up of Justices of the Peace who were responsible for regulating city clean-up, which included all yards, gardens, wharves and streets. All existing sanitary regulations were to be enforced at this time (Drolet 1967:70).

During the cholera epidemics, the local government re-established a local health office to oversee city clean up and to erect temporary hospitals. The epidemics provided the impetus to establish municipal boards of health in Québec City in 1833, 1840, 1849, and 1851. Evidence from 1856 suggests it was to be one of the permanent committees of the city and was to be re-nominated every year (Conseil de ville de la cité de Québec 1856), although this did not in fact occur, and there was no permanent committee until 1890 (Dagneau et al. 1983:153).

From 1832-1847, during fifteen years of horrifying, periodic outbreaks, there had been little change in the cleanliness of the city or the availability of services. Adding to these epidemics, was the deplorable state of the city and the financial and political indifference of the local politicians (Dagneau et al. 1983:154). The long term effects of the cholera epidemics were the cause for only a few lasting sanitary improvements which included the ban on inhumations within the city instated in 1855, and the 1858 government funding made available to hospital dispensaries who kept reserves of vaccines (Goulet and Paradis 1992:229). Other than these measures it appears there was little improvement in the sanitary state of the city in the first half of the nineteenth century.

While the second half of the century was marked by repeated waves of cholera and a smallpox epidemic in 1885, this period was also one that better served the public through the establishment of water and sewer networks, and the passing of stricter sanitary legislation. A permanent municipal health office was created, as well as a provincial governing body, the Conseil d'hygiène de la province de Québec (CHPQ), a result of the British North America Act of 1867, Article 92, which made health and social services a provincial responsibility (Goulet and Paradis 1992:235).

The municipal aqueduct

The construction of the municipal aqueduct has been considered by some as the major municipal works project of the second half of the nineteenth century (Hare et al. 1987:294). Poor water distribution was a problem and in 1842, James Gibb, John Munn and John Jones obtained the charter for the Québec Gas and Water Company. They were to construct a water system for the city and install twenty fire hydrants to it (Drolet 1967:27). Due to their inactivity, the charter was revoked three years later and the city undertook the project. In 1847 the City won the exclusive right to build and administer its water system to a distance of twenty-five miles, and the power to impose a tax once service was available (Drolet 1967:27). The services of George R. Baldwin, an engineer from Boston, were used, and he also oversaw the installation of gas lighting which was inaugurated November 1, 1849 (Hare et al. 1987:295-6). Previous to this, water was taken directly from the St. Lawrence River and distributed by water carriers or taken from wells (Laframboise 1991:79).

The water was taken from Lac Saint-Charles near Québec City. The canals reached the edges of the city in 1852, but the need for a sewer system slowed the project down and raised costs (Drolet 1967:29). The city aqueduct was inaugurated in 1854 and, somewhat ironically, may have encouraged the spread of cholera that year, but people were convinced that their standard of living was improved with piped in water (Drolet 1967:77). Shortly after the installation of the aqueduct began, it was realised a sewer system would have to be twinned with the aqueduct to properly service the city (Rapport de l'inspecteur des chemins 1852:3). Otherwise, drains for water would end up serving as sewers. Building the two systems in tandem ultimately caused a huge municipal deficit for many years (Drolet 1967:28).

Already, in 1854, there was some concern for the quality of water available. A Dr. LaRue was invited to analyse the water from the city aqueduct as it was suspected to be a factor in the spread of cholera (Dagneau et al. 1983:155; Goulet and Paradis 1992:234). His results stated that the water was full of strange substances with a foul odour and needed to be filtered, although this recommendation was not followed (Dagneau et al. 1983:155).

By 1863 the whole system had cost over a million dollars and consisted of a network of 147,353 feet of water pipes and 97,861 feet of sewers serving 3266 buildings. There were however, still 2000 houses that used the services of water carriers (Hare et al. 1987:296). In theory, there was enough water to serve three or four times the population of the city, but the loss from defective pipes, breaks in the system, and industrial abuse was immense. One brewery used 40,000 gallons a day, although they were billed at 6,000 gallons per day until the installation of counters revealed otherwise (Drolet 1967:29). It was soon apparent that the water was not enough to meet the demands, as Upper Town only had water from 8 a.m. to 8 p.m., and sometimes for only four hours a day, resulting in the use of domestic water reservoirs, while Lower Town generally had water for around twenty hours of the day (Drolet 1967:30). Less than ten years after the inauguration of the aqueduct, another main conduit was installed which cost almost half a million dollars (Drolet 1967:30). While not perfect, the system was an improvement in the city and people were no longer forced to drink water from the St. Lawrence River (Hare et al. 1987:297). According to Drolet, it was considered one of the best systems in North America and contributed to the reputation of the city, as did the sewer system (Drolet 1967:30).

An 1866 by-law stated there was to be no wasting of water and no sharing by those who paid for the service. Houses connected to the system that kept horses and cows paid an additional annual tax of one dollar. All houses connected had to pay for and have meters installed, which they rented annually (Chouinard 1901:88-89 [1866]). The City Engineer was given full access to any of these homes between 9 a.m.-9 p.m. for inspections (Chouinard 1901:90 [1866]), likely to ensure that people complied with the regulations.

Inconsistency in service appeared to have been an ongoing problem as a city by-law was passed in 1883 to improve water, ensure its constancy, and have sufficient pressure for fire pumps (Chouinard 1901:88 [1883]). By 1870, the aqueduct covered about half the streets of the city (Goulet and Paradis 1992:228), and in 1885 all streets in Québec had fairly efficient access to water (Dagneau et al. 1983:109). Conflicting data comes from a Provincial Hygiene Inspector's report in 1888. Dr. Joseph Beaudry commented that there was still no water available in three-quarters of the city, a common occurrence due to deficiencies in the piping of the water (ANQ, Rapport du CHPQ, 1888). This may either be erroneous reporting or perhaps the city was equipped with the networks although only a quarter of the population in 1888 was actually connected, due to their economic situation. In 1912, the engineer Lafrenière stated the polluted river where the municipal aqueduct was fed from had been a problem since 1893. He claimed there was no difference between water used to fight fires and that for drinking, and there were no means to prevent against infectious diseases of any nature that come from water contamination (Larocque 1970:46).

The installation of sewers and the privy pit menace

A functioning sewer network was needed for the city. As more buildings were constructed, the evacuation of used waters and human wastes needed to be addressed. Other than using privy pits, the normal means of disposal was to build ditches along the sides of streets, some of which were closed over (Drolet 1965:101). These were located next to the sidewalks and appeared to be about a foot wide (Chouinard 1901:96 [1845]). Buildings were then to be connected to these by drains. An 1823 report on these ditches on de la Fabrique street showed them to be the cause of frequent flooding into nearby cellars (Drolet 1965:101). Installation of such drainage ditches was done according to need and not part of an overall municipal plan (Drolet 1965:102). In 1849 the city road inspector stated the city needed a better drainage system (Goulet and Paradis 1992:228). In his report, Joseph Hamel advocated for 'canalisation' or channelling in a complete and scientific manner to connect water from houses, from street drains, and from privy areas, therefore meeting the needs of every home (Bureau de santé, canalisation de la ville 1849:4). He pointed out that on some streets the drainage was so bad, that the streets were not dry for even six months of the year, and many of the existing canals needed to be to be deepened. In 1853 it was decided to build a sewer network paralleling the municipal aqueduct and by 1854 26,000 feet of buried sewer pipe had been laid. By 1863, 97,861 feet of sewers had been built, 19,760 of which were constructed of brick (Drolet 1967:29).

The sewer network was not as efficient as the aqueduct; the pipes were too small and were frequently blocked as used domestic water, wastes, and rainwater were all drained by the same pipes (Dagneau et al. 1983:114). Around this time the road inspector's report commented on the anticipated reaction to the availability of the forthcoming sewer and water services. He stated that certainly people would want to install indoor plumbing, once water was introduced, but it would not be feasible; therefore the system would need supervision to avoid clogs, and freezing during the winter (Rapport de l'inspecteur des chemins 1852:10). Ironically, progress in the form of indoor plumbing had created a sewage crisis in many North American cities with leaky ineffective systems contaminating water and dampening basements (Tomes 1990:518-519). This convinced sanitary reformers that these modern conveniences were health hazards (Tomes 1990:519), and many people did not install indoor plumbing for this reason. Its presence would have conflicted with beliefs about the harmful effects of noxious odours. In Québec City in 1888, Beaudry notes that the most common system of indoor toilet is the pan-closet (ANQ, Rapport du CHPQ, 1888), which involved no indoor plumbing and had to be emptied manually. In 1891 Beaudry stated that the city did not have a complete sewer system, and that the present one was polluting the Saint-Charles River (ANQ, Rapport du CHPQ, 1891).

Despite the city having a first class aqueduct, Beaudry commented that there were very few properly installed water-closets in the city, which also lacked a main sewer collector. Instead, seven or eight pipes deposited sewage on the banks of the Saint- Charles River at low tide. At high tide it contaminated the streams and water sources. The sewers were proclaimed to be a greater hazard than a benefit as they were merely an extension of the privy system, due to citizens connecting their privies to the sewers which then become blocked as there was not enough liquid for proper drainage (ANQ, Rapport du CHPQ, 1891).

These water and sewer systems would have worked better if the sewer systems had preceded aqueduct construction, but this was never the case in North American cities (Duffy 1990:77). The ideal infrastructure would have all privies and cesspools draining into the sewers, but this had to be done properly to prevent backups. Drainage systems were to carry away surface water, and not the contents of privies and cesspools, a common problem (Duffy 1990:89). The installation of the water systems often strained the existing networks, which was worsened by the addition of flush toilets (Duffy 1990:89). In fact, the technology and design of the sewers lagged about fifty years behind that of the water networks (Duffy 1990:89). Due to ongoing problems with their system, Québec City employed a sewer collector up until the 1940s (Dagneau et al. 1983:114).

While the infrastructure was being constructed to serve as the city's sewer system, another sanitary problem was the presence of the thousands of privy pits in the city. Heralding from an era of the miasmic theory of disease and fear of noxious gases, the privy pit would have been a focal point of sanitary reform, representing the foul and the odorous as these represented danger, illness and even death (Paquet 1990:34).

Privy pits were waste receptacles, intended to receive human wastes, but were also regularly filled with other sorts of domestic trash. They could be simple holes built in cellars or yards, or adjacent to homes, or they were lined with a barrel or a wooden or brick structure. Outhouses were, at times, placed directly over the pits, while other homes connected the pit to their houses by drains, and had their facilities indoors. In Québec City in the first half of the nineteenth century, people connected these systems to the public canals (Laframboise 1991:182).

Québec City's 1866 by-law No.198 tried to standardise the construction of privy pits, as the city was still lacking a good sewer network. This by-law would serve as the guideline for sanitary officers until 1893. It required that the pits were dug at least three feet deep and connected to a drainage system if available. The pits were to be cleaned out only at night and at least once a year (Dagneau et al. 1983:156). That same year, anticipating another wave of cholera, city health officers covered over 492 privies while emptying 312 of them. They indicated that it was easier to implement this new by-law in the Saint-Louis and Palace wards where sewer channels and the aqueduct were already installed (Dagneau et al. 1983:156). One of the inspectors, Charles Darveau, pointed out that many people did not conform to the by-law unless they were forced. This by-law also stated that household garbage and animal carcasses were to be deposited at Pointe-aux-Lièvres, which was the main city dump until the twentieth century (Dagneau et al. 1983:156). A sewer tax was also created in 1866, indicating at least some houses were using this system (Hare et al. 1987:299).

In 1891, Inspector Beaudry stated that for Québec City, by far the most common method of human waste disposal was the privy system. He claimed privies were poorly situated, shoddily constructed, and under-serviced. In essence they were often just holes in the ground, regularly contaminated the surrounding soil (ANQ, Rapport du CHPQ, 1891). He attributed the permanent presence in the city of contagious diseases such as diphtheria and typhoid, to these poor hygienic conditions in which the population constantly lived (ANQ, Rapport du CHPQ, 1891). He also stated that pan-closets should be phased out as should all the existing latrines, as he advocated only the use of flushing toilets (ANQ, Rapport du CHPQ, 1891). Beaudry noted that the sewers in the city were often blocked and functioned so poorly that they did more harm than good. According to Beaudry privy pits must be at least 15 feet from the nearest building and they, and any cesspools, must be cleaned at least once a year and should not be connected with public drains (ANQ, Rapport du CHPQ, 1891). Even when water and/or sewage hook-up was available, people were often hesitant to put the old privies out of use. Archaeological remains from privy pits in the Greenwich Mews area of New York City indicate they were still in use in 1880 even though water hook-ups had been available since 1853. This was likely due to the cost of the hook up and the ensuing taxes (Geismar 1993:63).

As they were considered to be such a public nuisance and a hindrance to the good health of citizens, privy pits were a focal point of the municipal Board of Health's attempt at sanitary reform. For the year 1892, their principal projects were the regulation of waste pipes and drains in the interior of houses, the suppression of privies and defective water-closets, and the elimination of wooden drains connecting to the public sewers (Archives de la ville de Québec [AVQ], Rapport des opérations du bureau d'hygiène de la Cité de Québec pour l'année 1892, QC3-2E/554-2).

Complaints to the Board of Health were also received on this matter;
"Le médecin municipal a attiré l'attention du comité sur certaines conditions qu'il considère non-hygiéniques et qui sont toutefois encore tolérées dans la ville parce qu'il n'y a pas de règlement municipal qui permet au bureau d'hygiène de les faire disparaître... aux fosses d'aisances, aux latrines... et à certains water-closets de construction défectueuse" (AVQ, Minutes, Bureau d'hygiène, 1892, QC3-1B/538).

The following year, 1893, the city passed another sanitary by-law which stated that material cleaned from the privy structures, which were to be cleaned at least once a year, was to be transported at night in sealed containers, and might only be done once the necessary permit had been acquired. Privy contents were not to be transported in the same vehicle as food items, and the responsibility for the annual disinfecting of the pits fell upon the owner or occupant of the home. When the occupants were not connected to the city aqueduct and drainage system, the pits were to be at least three feet deep, and, if possible, a minimum of fifteen feet from dwellings. The outhouse over or associated with this privy had to be a covered building at least six feet high. These systems were not to be connected with the public sewer system and any existing connections were to be closed off before the 1st of November 1893. All homes on streets with sewer connections available were to install water-closets, connect them to the sewers, and have all privies out of use by the 1st of May 1895. The remaining pits were to be completely emptied and filled with earth (Chouinard 1901:118-122 [1893]).

According to the 1893 by-law, the city attempted to eliminate the use of privies by 1895, (Chouinard 1901:121 [1893]) an ambitious undertaking as the annual report of the Board of Health for this year cited 2282 privies within the City limits (AVQ, Rapport des opérations du bureau d'hygiène, 1893 and 1894, QC3-2E/554). While the municipal Board of Health's annual reports indicate that every year more water-closets were installed and privies put out of use, many people also requested extensions to the 1895 deadline, indicating that the city was not entirely effective in ridding itself of this problem. The 1895 amendments to the Public Health Act of Québec (Article 3058) state that the Provincial Hygiene Council was to eliminate all nuisances or causes of illnesses, especially the construction and the maintenance of public sewers, privies, drains, and cesspools, and to maintain or better the public health (Rapport du conseil d'hygiène de la Province de Québec 1895).

After creating the by-law of 1893, the number of privy pits within the city decreased annually but at a much slower rate than had been foreseen. The inability of the Board of Health to upkeep the by-law is seen in the yearly requests presented to the Board of Health for extensions on deadline. In 1896, 357 people were granted a one-year extension (AVQ, Rapport des opérations du bureau d'hygiène 1896, QC3-2E/554) and in 1898, 471 more allowances were granted, (AVQ, Rapport des opérations du bureau d'hygiène 1898, QC3-2E/554) as were an additional 147 in 1900. At that time the board claimed there were only 348 privies in the city

(AVQ, Rapport des opérations du bureau d'hygiène 1900, QC3-2E/554). In considering their figures, 369 privies remain unaccounted for, and even after the deadline to suppress all privies, annual reports indicate new privies being built according to municipal regulations. In 1908 the Conseil d'hygiène de la province de Québec (CHPQ) reported there were still 250 functioning privies within the city which, when informed, the municipal doctor, Dr. Catellier, refused to believe (Larocque 1970:44).

The pan-closet systems were banned in any new buildings by the Provincial Hygiene Council (Quebec Official Gazette 1891:1479-1490) and the only system allowed was the *chasse-d'eau* or flush toilets (Chouinard 1901:123 [1893]). All new buildings were to be approved by the City Engineer to assure compliance with the public's health and security (Chouinard 1901:123 [1893]). The annual reports of the Board of Health indicate progress in the installation of water-closets as there were 4,483 in the city in 1891-1892, a number that jumped to 11,370 by 1900-1901 (AVQ, 1900, QC3-2E/554).

Street Maintenance

After the installation of the water and sewer pipes, many city streets needed repaving. The city added 11,256 feet of sidewalks during the years 1855-60 (Hare et al. 1987:297). At that time the paving options were either small stones, wood planks, pine or red spruce blocks, or macadamising, which was the laying of uniform sized stone blocks (Drolet 1967:36). Even though they were not the recommended forms, in mid-century the city continued to pave with the macadam method and in wood, particularly for the sidewalks, which had to be recovered soon after (Drolet 1967:37).

Winter street maintenance was the responsibility of property owners. This was a heavy burden for the city to bear as they received only voluntary payments from citizens and some government funding (Drolet 1967:37). Joseph Hamel, the road inspector tried to regulate this publicly and a tax was to be paid according to the property value and by number of feet to be cleared (Drolet 1967:37). As of 1866, people paid for the half of the width of the street, and the fees were shared accordingly amongst the property owners (Drolet 1967:37).

According the provincial by-laws of 1891, garbage was to be removed at least once a week from 15th April to 15th October. On 1st of May, all cellars were to be cleaned out and any animal or vegetable matter removed. Public dumping was to be at least 1200 feet from any dwelling or any wells from which drinking water was obtained. Where the Municipal Council did not provide for garbage removal, the head of the household was to bury or burn it at least one foot underground, unless it was used for manure. Even then it was to be covered in ashes and 100 feet from any source of drinking water (Quebec Official Gazette 1891:1479-1490).

That same year, the Provincial Hygiene Inspector Joseph Beaudry observed that there was no regular garbage removal and it was left to private initiatives. He complained that although citizens were to clean their yards, they did not do so, and they threw everything into their privies or onto vacant lots. At times, farmers came into the city and took stable and animal wastes for fertiliser, but this was not done regularly. He also mentioned that it was critical for a large city like Québec to have a proper garbage collection service, and this service should be under the city's jurisdiction, as it was the only means to ensure that it would be done. He recommended compulsory garbage removal every eight to fifteen days according to the quantity produced (ANQ, Rapport du CHPQ, 1891). When garbage was, in fact, removed it was often taken to Pointe-aux-Lièvres, on the Saint-Charles River, which served as the municipal dump. The Provincial Hygiene Inspector, Dr. Joseph Beaudry, made two inspections of this depot in 1893 and 1898.

The Board of Health annual report for 1893 and 1894, states that the one accomplishment not made by their office that year was the organisation of the removal of city garbage (AVQ, Rapport des opérations du bureau d'hygiène, 1892 et 1893-1894, QC3-2E/554-2). The board members considered decomposing substances a hotbed for the propagation of disease germs. The simplest means of garbage disposal was for people to burn them in their kitchen stoves, but most people, instead, allowed them to accumulate in yards where they became a source of danger. When gas stoves replaced traditional ones, inhabitants allowed garbage to accumulate for some time before having it removed. The board members went on to state that they do in fact have an organised core of street cleaners, but daily garbage removal was needed in the summer as it presents a grave danger (AVQ, Rapport des opérations du bureau d'hygiène, 1892 et 1893-1894, QC3-2E/554-2). One of the main preoccupations of the Québec City Board of Health during the latter half of the nineteenth century was vaccination against smallpox (Dagneau et al. 1983:159). Duties of local boards of health included the inspections of properties for smallpox and general cleanliness (Quebec Official Gazette, 1885:1703).

As the by-law of 1893 stipulates, the cleaning of privies, streets, yards, dwellings and snow were all left to the personal responsibility of the owner or occupant of the dwelling (Chouinard 1901:112-129 [1893]). Annual reports from Québec City indicate the Board of Health was aware of this problem, and questioned why the City could not afford a service to clear these wastes away. It was one of the Board of Health's mandate, but was not achieved, and, in fact, a regular garbage service was not provided for another forty years (Dagneau et al. 1983:161).

Sanitary by-laws and the creation of the Conseil d'hygiène de la province de Québec

To combat another cholera epidemic in 1866, the city passed a sanitary by-law (Dagneau et al. 1983:155). No other new sanitary by-laws were passed by the city until 1893, but in the interim, in 1885, the province of Québec created a central health office, although it was not formally established until the following year (Goulet and Paradis 1992:249). In 1886, the Public Health Act was passed creating the Provincial Health Council (Bernier 1989:103), likely as a reaction to the devastating smallpox epidemic of the previous year. The Public Health act was amended in 1888, 1890, 1892, 1894, 1895, 1901, and 1915. In 1887, the Provincial Health Council or Conseil d'hygiène de la province de Québec (CHPQ) was officially formed (Guérard 1999:208). It had seven founding members, four of which were doctors. They studied contagious diseases, their prevention, oversaw the formation of local boards of health, created provincial by-laws, and dictated the policies and procedures for the reporting of contagious diseases (Goulet and Paradis 1992:249). In 1888 they demanded that all municipalities have permanent health offices (Dagneau et al. 1983:160). By 1895 there were 839 boards of health in the province out of a total of 876 municipalities (Rapport du conseil d'hygiène de la province de Québec 1895:21).

One of their goals was the creation of a municipal health code that would be applied equally, without exception (Guérard 1996:208). The 1895 amendments to the Public Health Act of Québec (Article 3058) state the CHPQ was to eliminate all nuisances or causes of illnesses, especially the construction and the maintenance of public sewers, privies, drains, and cesspools, and to maintain or better the public health (Rapport du conseil d'hygiène de la province de Québec 1895). The office was responsible for overseeing the enforcement of these rules by the municipal offices, although the secretary's report from 1895 states that often, some of the municipalities were apathetic to many sanitary measures. Local authorities resented the role of the CHPQ as they considered it a loss of their autonomy, and were uncertain as to whether they actually needed such a governing body or not (Guérard 1996:208). According to Goulet, the efforts of the CHPQ were likely met with some scepticism as many communities were divided on medical matters and they were not always comfortable with those proposed by the CHPQ (Goulet 1996:506).

The CHPQ inspectors were sent out to investigate possible epidemics, what they considered public nuisances, and to make sanitary inspections. Inspections were made of Québec City in 1888, 1889, 1891, 1893, and 1898 regarding a possible typhoid fever case, general sanitary inspections of the city, and inspection of the municipal depot or garbage dump, which was also categorised on a return visit as an inspection of a 'nuisance'. In general, they paint a bleak sanitary picture of the city.

According to his report based on a three-day inspection in November 1888, Dr. Joseph Beaudry stated the Board of Health existed in Québec City in name only. They had not, according to Beaudry, adapted any of the new scientific progresses in sanitation. A great number of streets were filled with garbage, and as privy pits were the most common structures in the city's small backyards, they were often situated very close to the neighbouring property. He concluded that, on public health matters, there was no protection for the population (ANQ, Rapport du CHPQ, 1888).

Visiting a year later, Elzéar Pelletier commented that for a city with a population of 80,000, the Board of Health rarely met, the Chief Health Officer had neither office nor inspectors, and the laws in place to combat the spread of contagious diseases went unenforced (ANQ, Rapport du CHPQ, 1889). On this brief visit to inspect cases of typhoid, Pelletier found the state of the city so deplorable he asked at once to meet with the Board of Health. He was horrified that the City Doctor did not know the source of the typhoid breakout, and correctly identified the source as a dairy in Saint-Sauveur who was continuing their daily milk deliveries (Dagneau et al. 1983:161). He then demanded a revision of The city's sanitary laws, regular collection of garbage, the connection of homes to the available public sewers, the abolition or at least regular cleaning of privies, and for the removal of the municipal tax on water-closets, a fee which has been in place since at least 1873 (Chouinard 1901:92 [1876]), to encourage their installation by all citizens. He was also appalled that garbage and dead animals were thrown into the Saint-Charles and St. Lawrence rivers, and slaughter-houses were allowed in places where the tides would later bring up their wastes (ANQ, Rapport du CHPQ, 1889). An emergency meeting of the Health Office resulted in the promise of new by-laws (Dagneau et al. 1983:161).

A more thorough inspection was made of the city in 1891 with disastrous results. Inspector Joseph Beaudry's scathing report, which was to remain confidential, was leaked to the press. The condemnation of the city's state was so severe, the City Council countered that the reports contained the fancy of an ill-informed person who did not have the authority to make such claims (Dagneau et al. 1983:161). The president of the Board of Health stated that parts of the report were incorrect and exaggerated (AVQ, Minutes du bureau d'hygiène, 1891, QC3-1B/538). This exaggeration and fancy of Beaudry's was a twenty page report detailing all aspects of the sanitary state of the City including slaughter-houses, industrial establishments, public markets, and buildings. Beaudry identified three main problems, which were a lack of municipal garbage collection, poor drainage and sewer systems, and general carelessness in the construction of cesspits (Dagneau et al. 1983:162). He found the local Board of Health to be inefficient and incapable of implementing current sanitary by-laws. The City responded by dismissing the municipal doctor, Dr. LaRue and replacing him with Dr. Catellier, who made new recommendations (Goulet and Paradis 1992:256). While these were new sound resolutions on dealing with epidemic diseases, they did not address Beaudry's major complaints (Dagneau et al. 1983:164).

Beaudry would visit the city again in 1893 on an investigation of a public nuisance. This nuisance was the city's garbage piled up at Pointe-aux-Lièvres. This strip of land on the Saint-Charles River, was the central receptacle for the city's garbage, which included household wastes, dead animal, and privy pit contents. The odorous mound according to Beaudry had reached a height of six to seven feet and he demanded the city take action (ANQ, Rapport du CHPQ, 1893). Five years later he was sent to inspect this same problem, as it was still the main city dump, and recommended that the City at least cover the material in a foot of earth to cut down on the odours, as outlined by the CHPQ (ANQ, Rapport du CHPQ, 1898).

Above and beyond the privy regulations which have already been outlined, the municipal 1893 sanitation by-law, *Concernant la propreté, la salubrité, et l'hygiène, dans la cité de Québec,* stated that all vacant lots within the city had to be enclosed by a four-foot fence. All properties were be kept tidy, without any garbage, trash or anything unhealthy, inconvenient or emitting a bad odour. It was the responsibility of renters and owners alike to clear off their properties and in front of their houses, and occupants could be fined for not doing so. No garbage was to be thrown into the streets, nor were rugs to be cleaned or liquids disposed of there. Snow must be removed before the 1st of May unless otherwise indicated by doctor's permission (Chouinard 1901:112-114 [1893]). No animal or vegetable material was to be left in cellars, thrown in privy pits, or yards. These materials had to be burned or be transported to the appropriate place determined by the City Engineer (Chouinard 1901:112-113 [1893]). The keeping of pigs was not allowed in the city, nor were fowls of any species allowed to be kept in domestic dwellings. The by-law also outlined the sales of meat, butter, and milk, the appropriate care and cleanliness of market stalls, the licensing of slaughter-houses, and appropriate butchering practices (Chouinard 1901:115-118 [1893]).

Clearly the dire situation at Québec, indicated by the inspection reports, required sanitary reform. Sanitary improvements, however, appear to have only been made when requested by the CHPQ, and even their recommendations were not always heeded (Dagneau et al. 1983:154). A law amending the power of the CHPQ interventions was passed in 1894. It stated that once founded, boards of health could not be abolished, and had the ability to pass local by-laws necessary for their proper functioning. Municipal authorities were also given the power to regulate industrial standards of hygiene (Goulet and Paradis 1992:259).

The Board of Health in Québec City began to function permanently only in 1890 (Dagneau et al. 1983:153). The municipal doctor was, a 'hygiéniste militant,' but many of his efforts would have been ineffective due to the financial and political indifference of the municipal authorities to the city's sanitary state (Dagneau et al. 1983:154). Beaudry's 1891 report indicated that the Board of Health appeared to exist only to satisfy the law requiring its presence, was the smallest department of City Hall, and was far from having any importance there (ANQ, Rapport du CHPQ 1891). He concluded the Board of Health office did not have enough money to make improvements, and was weak and powerless. Furthermore, the board did not respond to the public's needs, which undermined the public's confidence in its ability to be effective (ANQ, Rapport du CHPQ 1891).

While the image of the city from the CHPQ inspectors' reports portrays the Board of Health as inactive and indifferent to the plight of its citizens, evidence of the board's activities from 1891-1900 indicates somewhat of a different situation. Between 1892 and 1900, the board's two inspectors carried out almost 5,000 sanitary inspections annually and sent out over a thousand letters a year on sanitary issues and offences in the city for most years between 1890-1900 (AVQ, Resumé des opérations du bureau d'hygiène, 1891-1900, QC3-2E/554-1-4). They paid for the cleaning of a poor citizen's privy in 1892, and notes of their meetings indicate they were discussing and acting on problems with the sewer system, and were fully aware of the abundance of garbage and privy pits problems in the city (AVQ, Procès-verbaux du bureau d'hygiène, 24 avril 1890-11 avril 1916, QC3-1B/538).

The responsibility for creating and legislating acceptable standards of public sanitation during the end of the nineteenth century had become the realm of the CHPQ. Originally, the federal government had only intervened on matters such as quarantines and vaccinations (Goulet 1996:494). Now the province, via the CHPQ, was attempting to normalise and standardise hygienic reforms in the municipalities of the province, a process that was met with some resistance. The municipalities were afraid to relinquish their autonomy and resented the restrictions put upon them. Local authorities were also not always convinced of the need to implement all the new sanitary measures and represented a divided public (Guérard 1996:208). Furthermore, the imposition of these new measures demanded money from municipal budgets that were already straining to meet the public's needs (Guérard 1996:209). The CHPQ itself also lacked the funding, staff and the ability to enforce sanitary measures. They may have created provincial laws, but in reality did not have the ability to implement them (Guérard 1996:213). For example, Copp points out that the Québec Health Act was highly flawed, as the CHPQ could force a municipality to create local health boards, but no municipality could be forced to enforce the regulations or finance the programs (Copp 1978:91). In addition, the CHPQ encountered difficulties in uniformly applying sanitation laws, and the medical community and clergy were of dissenting opinion on medical and sanitary subjects such as vaccination and disease propagation (Duffy 1990:196; Goulet 1996:493). Starting in 1890, the Québec press, medical communities and labour unions began lobbying for sanitary changes such as improved municipal garbage collection, industrial safety, plumbing by-laws, and public bath houses (Dagneau et al. 1983:168-9), and were thereby united in promoting reforms. While there were many infrastructure improvements in the city in the second half of the nineteenth century, the CHPQ reports indicate a city that was ill equipped to deal with its sanitary problems.

CHAPTER THREE

THE ÎLOT HUNT SITE

3.1 HISTORY OF THE ÎLOT HUNT SITE

History of Lower Town, Québec City

Québec City has always been both geographically and culturally, a divided city. The Upper Town area (Appendix A, Figure 1, see insert) on top of the Cap-aux-Diamants cliff was the colony's seat of administration, religious and educational institutions. In the nineteenth century, this area continued to serve these functions and house the growing middle class neighbourhoods. The Lower Town area, a thin strip of land wedged between the Cap-aux-Diamants cliff and the St. Lawrence River was the site of Champlain's fort, Place Royale, a densely populated area of homes and businesses, the Canoterie cove and the Intendant's Palace. The Lower Town, including Place-Royale and the area directly north of it, became the economic hub of the city, and was closely tied to the port industries of trade and shipbuilding.

The population of the city grew from 28 to 7215 during the French regime (1608-1759) (Desloges 1991:21-23). In Lower Town, the mostly stone buildings covered two thirds of the surface area during the 1700s (Desloges 1991:87), indicating a dense urban landscape. After destruction during the British Conquest of 1759, this area of the city was rebuilt, and by 1795, there was an increase in the number of houses which were primarily two or three-storey structures (Hare et al. 1987:120). During the eighteenth century, Lower Town business proprietors normally lived above their places of work (Desloges 1991:51), and the success of these residents is indicated by the fact that this was the sector of Québec City with the highest amount of domestic help (Desloges 1991:57).

By 1791, there were 13 wharves in the port area, which prospered in part due to the war in the southern Colonies (Hare et al. 1987:120), and by 1804 the number of wharves had grown to more than 20 (Hare et al. 1987:150). Also in the early nineteenth century, the banking industry became established in this area, with most of the institutions housed on rue Saint-Pierre, the location of many of the city's commercial activities. In 1818, the old wards of the city housed half the city's population, though in 1861 they made up less than a third, a testament to the rapid expansion into the newer areas such as Saint-Jean (Hare et al. 1987:212-213).

By 1820, Québec City was considered the most important Colonial centre in North America due to its importance in the timber trade, furnishing the Royal Navy with squared timbers (Ruddel 1981:3). When cholera struck in the 1830s, the port area and the adjacent Saint-Pierre ward (Appendix A, Figure 6) were badly affected. As the city received 600-10,000 immigrants weekly during the summer shipping season, diseases spread quickly in these crowded areas (Bilson 1980:7) (Appendix A, Figure 7). The demographic profile of the Saint-Pierre ward now shifted as many of the middle class inhabitants of the area moved into the Upper Town (Hare et al. 1987:213), and were replaced by Irish immigrants.

As Lower Town slowly expanded by reclaiming land along the river's edge, the port doubled in size by the mid-nineteenth century, wharves were extended and a third storey was added to the many warehouses (Hare et al. 1987:214). Québec, Upper Canada and the States all sent wood to Québec City for export (Ruddel 1981:250). Québec also was exporting wheat, potatoes, and maple sugar, though wood was the key industry in the developing shipbuilding trade (Laframboise 1991:10). The Saint-Pierre and Saint-Louis wards, around 1840 had mostly trades people and workers employed in boat building and shop keeping (Drolet 1967:99). By the 1850s, the most valuable properties in the city were the stores, businesses, warehouses, and hotels of Lower Town (Drolet 1967:99).

From 27,741 in 1831, Québec City's population had more than doubled in 1871 to 59,699 (Drolet 1967:100). By the mid-1800s, Montréal began to eclipse Québec City and grew rapidly for the next fifty years, from a mid-century population of 57,715 to 267,730 by 1901. During the same time period, Québec City increased from 42,052 to 68,840 (Ruddel 1981:67), ceasing to grow as a city around 1861.

This was due to many reasons which included the choice of Ottawa as the permanent national capital, Québec City's diminishing role as a wood provider, and the growth of the railway, which reduced the need for water transport (Drolet 1967:102). While there was some industrial expansion that replaced the shipping trade during the last quarter of the century, Québec City did not experience the tremendous late century growth of cities like Montréal and Toronto.

From 1843-1871, the population in the Saint-Pierre ward was relatively stable and averaged around 3,600 people (Drolet 1967:104). The whole Lower Town area made up around 20 percent of the city's population, while Saint-Jean and Saint-Roch contained 70 percent of the inhabitants and three quarters of the buildings (Hare et al. 1987:278-9). In the 1870s, the population and character of the city's sectors were clearly divided on a social and economic hierarchy, with the merchants, professionals and civil servants in the upper city, and with the workers in Lower Town.

The 1870s were the beginning of a period of economic and demographic stagnation as the wood construction industry declined, coupled with the departure of both the city's civil servants and the British army regiment (Dagneau et al. 1983:154). This decline in ship building, and the switch to steam eventually eliminated the local ship building industry (Dagneau et al. 1983:35), an enterprise which was central to the economy of the port area of Lower Town. The Saint-Pierre ward experienced a reduction in its population during this period from 3727 in 1871, 2480 in 1881, to 2712 in 1901 (Dagneau et al. 1983:39), as some people moved closer to the industrial ward of Saint-Roch.

The History of the Îlot Hunt Site (CeEt-110), Lower Town

The Îlot Hunt site (CeEt-110) is contained in a rectangle of land bordered at its western edge by rue Saint Pierre, rue Saint-Antoine, rue Dalhousie to the east, and cadastral lots 2120 and 2120-3 to the south. The Îlot Hunt site itself is made up of cadastral lots 2118-4-12 (Appendix A, Figure 8).

The land was first conceded in 1687 to the merchant Charles Aubert de la Chesnaye, who built a wooden wharf on the property (Leclerc 1998:18). This was later replaced with a stone one in 1699, a structure that also contained a cistern which provided water for a bakery on the property (L'Anglais 1993:21). Rue Saint-Pierre was already established at this time as it dates back to 1661-1663, the period of Governor Davaugour (Desloges 1991:71), while rue Saint-Antoine was just a passage to the riverbank and not a recognised city street (Simoneau and Rouleau 1988:84).

The six cannon Dauphine Battery was constructed in 1707-1709 (Appendix A, Figure 9) connecting to de la Chesnaye's 1699 wharf (L'Anglais 1993:32). The ground below this battery which, had been constructed directly on the beach, immediately became the site of refuse deposits. Confirmation of this was found in Saucier's (1958) thesis on health and sanitation during the French Regime. An ordinance dating to April 16th, 1710 obliges all the inhabitants of the Lower Town; "de porter ou faire voiturer toutes leurs ordures, vidanges et démolitions dans l'emplacement de feu sieur Aubert de la Chesnaye dont on a tiré la terre pour faire des batteries" (Saucier 1958:61). Later, in 1750, the garbage from Lower Town was to be dumped at the end of rue Saint-Pierre (Saucier 1958:62), in order to create land. In 1721 the architect Joseph Maillou acquired the property. Between 1722-27 he built the 'Maison Maillou', later referred to as the 'Hunt house', on the corner of rues Saint-Pierre and Saint-Antoine (de Varennes 1994:7). This house was three storeys high with a cellar and an adjacent storehouse (de Varennes 1994:8).

Due to the War of Succession in Europe (1740-1748), and the siege of Louisbourg in 1745, a second battery was built in 1745 some metres to the east of the first one (Leclerc 1998:22), and the land lying between the two defensive works evolved into the Îlot Hunt site. The presence of the military batteries explains the slow growth in the site's evolution in comparison with the adjacent properties during this period (Le Groupe Harcart 1989:6).

The Maison Maillou and other buildings on the property were destroyed during the siege of Québec. Stephen Moore and Hugh Finlay built a new house in 1765 after they had acquired the land in 1763 (Leclerc 1998:31). In 1783, there was also a long building running east to west along rue Saint-Antoine to the eastern end of the property, approximately the location of the second Dauphine Battery (Leclerc 1998:24). Charles Ward Althorp acquired the land in 1795, who in turn sold the land to the cooper John Chillas (Simoneau 1987:6). Chillas, five years later, was granted the waterfront lot at the eastern edge of his property, east of the second Dauphine Battery, and was under obligation to build at least one wharf and some buildings (Leclerc 1998:24). His notarial records indicate that he had a contract in 1797 to furnish the British Army with "500 powder barrels of well seasoned oak" (Ruddel 1981:252), indicating his prosperity. At the turn of the century, the property contained the reconstructed house at the corner of rues Saint-Pierre and Saint-Antoine, which also had a storehouse and a storey and a half addition on its southern side. The other buildings on the property were likely constructed of wood, and were stables or sheds, while the building that ran along rue Saint-Antoine appears to have been shortened on its eastern edge (Leclerc 1998:25). The inventory records made at his death describe a stable and coach house, and describe the property as having wharves, stores, houses and other buildings (Les Recherches Arkhis 1990:170-171). At this time, the second Dauphine Battery was used as a wharf.

Chillas developed his purchase into a busy nineteenth-century site of urban enterprise. He and John Goudie enlarged his existing wharf around 1815 (Drolet 1967:127; Leclerc 1998:25). Chillas also constructed a three-storey house on rue Saint-Antoine in or around 1815 (L'Anglais 1993:51; Simoneau 1987:60) (Appendix A, Figures 10, 11). Adjacent and to the east, was a second house constructed in 1824 (Leclerc 1992:25) (Appendix A, Figures 11, 12), later divided into two houses around the end of the 1800's (Dubé 1995:54). He also constructed two warehouses around this time, behind the Maison Maillou (Leclerc 1998:25)

(Appendix A, Figure 13). Chillas had been conceded another eastern addition to his property on the riverfront, and constructed a warehouse in 1822 along rue Saint-Antoine (Simoneau and Rouleau 1988:84) which was renovated, and today houses the Auberge Saint-Antoine. The southwest corner of the building was rounded off, presumably to facilitate traffic flow into the yard area behind (Leclerc 1998:25). It is unknown when the previous hangar, which existed in the late 1700s, was dismantled for the construction of the 1822 building.

Chillas died in 1825, leaving a third each of his estate to his children Elizabeth and George, while the remaining third was divided between his other children. Ten years later Elizabeth and her husband Thomas Hunt had bought the entire property from the other heirs, and were granted an additional easterly parcel of land (Leclerc 1998:26). At his death, Thomas Hunt left all the property to Elizabeth who renounced her legacy, and by 1838 Thomas's brother James, a master sailmaker, purchased all of Îlot Hunt (Leclerc 1998:26). This sale describes the 'Hunt house' on the corner of rues Saint-Pierre and Saint-Antoine, with two stone stores behind it, two houses on Saint-Antoine, stone offices, and various wooden buildings on south side of the lot, as well as a large wharf and a large stone store facing onto rue Saint-Antoine (Leclerc 1998:27; Simoneau 1987:7). James Hunt was a successful businessman in the city; he owned several other properties in the city, and sat on the board of the Banque du Québec from 1825-1830 (Bervin 1991:133-134).

During 1830-1850, rue Saint-Antoine was improved (Appendix A, Figure 14) as municipal authorities erected a wharf that extended the length of the street out to the low tide level. It had a public well, pumps and a staircase at the eastern end to provide public access to the river (Laframboise 1991:85; Simoneau and Rouleau 1988:85, 93).

By 1845, an additional building had been constructed, and the site attained its mid- nineteenth century character, which was a mixture of commercial and domestic dwellings (Appendix A, Figure 15). The house constructed in 1815 on rue Saint-Antoine had its southwest corner rounded off, as did the warehouse situated to the east. This was likely due to the volume of traffic on the site, as it provided easier access to the yard for horse-drawn carts. The site also contained a bakery in the southwest corner of the property, as well a stable, and other buildings (Leclerc 1998:27), which were presumably storage buildings or sheds related to the site's commercial activities. James Hunt died in 1847, leaving the property to his daughters Henrietta, Eliza, Mary and Amelia (Leclerc 1998:28)

Henrietta and her husband, Weston Hunt, had their business in the 1822 stone warehouse on rue Saint-Antoine (Simoneau 1987:8). They were wine merchants, and later owned a general store in the building, under the names of Weston Hunt & Co., and later Hunt Brock & Co. (Leclerc 1998:28). In 1871, the Hunt inheritors built a new warehouse on the eastern end of the wharf near the staircase to the water (Leclerc 1998:28). At the same time there were two other warehouses south of the 1822 warehouse (Leclerc 1998:28), indicating the density of commercial activity on the site.

The extension of rue Dalhousie in 1875 altered the warehouse on rue Saint-Antoine, as well as the two buildings directly south of it. All were literally carved in two by the construction of this new street (Appendix A, Figure 16). From then on, the transfer and shipping of merchandise was done at the end of the wharves in the large storehouses (Simoneau and Rouleau 1988:85). According to the 1879 plans, the houses on rue Saint-Antoine were extended easterly around this time, to connect with the store (Leclerc 1998:29). The two rectangular buildings behind the Hunt houses were also three storeys high, one of which was stone while the other, a stable, was made of wood (Leclerc 1998:29). The house on rue Saint-Pierre, south of the Hunt house, was a hardware store, and the three buildings along the southern wall delimiting the property were storage buildings (Leclerc 1998:29).

The site changed little until the mid-twentieth century. By 1957, the three-storey stone building behind the Hunt house and the adjacent stable were torn down. The houses along rue Saint-Antoine, and in the backyard were destroyed in 1960 (Dubé 1995:60). The wooden buildings along the southern wall of the site were destroyed in the late 1980s (Leclerc 1998:29), leaving only the Hunt house at 66 rue Saint-Pierre, and the stone building directly south of it at 58 Saint-Pierre at the western edge of the site. Îlot Hunt was purchased in 1990 by Immeubles de la Batterie, Inc., and the building at the corner of Dalhousie and Saint-Antoine was restored, while the other buildings facing Dalhousie were demolished. Restorations on the Hunt house began in 1995 and it is currently part of the Auberge Saint-Antoine. The former yard area between the hotel and the Hunt House is currently used as a parking lot for the patrons of the Auberge Saint-Antoine.

Îlot Hunt investigations 1987-1994

Îlot Hunt has been the subject of several investigations (Bain 1997; Bouchard 1994,1998; Boucher 1999; Cloutier 1991; de Varennes 1994; Dubé 1995; Fortin 1996; Goyette 1994, 1999; Lafleur 1999; L'Anglais 1993, 1998; Leclerc 1992, 1998; Le Groupe Harcart 1989; Moussette and Auger 1997; Simoneau 1987). The results of the archaeological excavations between 1991-1994, under the supervision of Marcel Moussette, will be briefly outlined (Appendix A, Figure 17). The 1995 excavation results, the source of much of the environmental and material data used in this project, will be discussed in the following section.

After studies of the archaeological and historical potential of the site (Le Groupe Harcart 1989: Simoneau 1987), Université Laval began excavations in 1991. The first Dauphine Battery (Leclerc 1998:59) was identified as well as a perpendicular wall thought to connect to the second Dauphine Battery (Leclerc 1998:61) (Appendix A, Figure 18). These two defensive structures were used as part of the foundations of 1815 and 1824 houses situated on rue Saint-Antoine, also located during the excavation (Leclerc 1998:65-66).

The following year, a caisson or stone box-like structure (Appendix A, Figure 18) was located in the 1687 de la Chesnaye wharf, which was 6 m high and 1.6 m wide, as well as a cistern dating to around 1700 (L'Anglais 1998:21-28). A privy (10A10), which was constructed around 1825, was also identified (L'Anglais 1998:89) (Appendix A, Figure 18). The same privy was found and sampled during the 1995 investigations. A hangar or storage building dating to the second quarter of the nineteenth century was also located (L'Anglais 1998:96).

The 1993 excavation was focussed on the area south of the houses on rue Saint-Antoine. A wooden drain (11C19), dating to around 1825, was excavated behind the house built in 1824, and ran diagonally in a southeastern direction (Goyette 1994:48). A second drain (11C11) (Appendix A, Figure 4) dating to around 1850 was also discovered, with virtually the same location and orientation to the first one (Goyette 1994:50). A large L-shaped privy structure (11A64, 11C15/11C35, 11B47, 16D200/16C26) was also discovered (Goyette 1994:51) (Appendix A, Figures 2, 18). One section of the privy was parallel to the Dauphine Battery while the other was perpendicular. A covered drain, at the easternmost end of this structure, was also discovered and excavated (11B50/55) (Goyette 1994:56) (Appendix A, Figure 19). It is this privy that became one of the focuses of the 1995 Îlot Hunt excavations (16D200/16C26). Remains were also identified of a building (Appendix A, Figure 20), that corresponds in location to one on 1875 insurance maps (Goyette 1994:61).

In 1994, the house foundations and cellar of the 1815 rue Saint-Antoine house were uncovered (Dubé 1995:22). The house immediately to the east, dating to 1824, was also located, along with a circular stone structure thought to be a well, which was constructed in the foundations at same time (13B23) (Dubé 1995:53) (Appendix A, Figure 18). The 1824 house was divided into two and a cellar was built around the end of the nineteenth century. Evidence of this alteration was located along with a sewer pipe (13B15) marked "W. D. Bell, Québec" dating to the late nineteenth-century, and a gas line (Dubé 1995:47, 55-56).

3.2 1995 EXCAVATIONS AT ÎLOT HUNT

The 1995 excavations of the Îlot Hunt site were undertaken by Université Laval's annual historical archaeology field school, under the supervision of Réginald Auger and Philippe de Varennes. As several interventions were planned for this, the final season of the Îlot Hunt excavations, the dig members were divided into three groups, while Philippe de Varennes carried out survey work on the Hunt house (Surveys 16A and 16B) (Appendix A, Figure 17), which was undergoing renovations. The team I supervised excavated sub-operation 16D (Appendix A, Figure 17). Our objectives included the excavation of the 16D200 privy structure and of the area east of the Dauphine Battery (16D100/16K100), with the goal of reaching the early eighteenth century beach level. Areas 16K and 16M (Appendix A, Figure 17) were extensions of this sub-operation, and were carried out by Claude Lafleur and her team.

This group also excavated sub-operation16E, which was extended to sub-operation 16F (Appendix A, Figure 17). They focussed on the house foundations along rue Saint-Antoine with the aim of excavating the circular stone structure found in 1994. This team also worked on the Hunt house survey in areas, 16G, and 16J (Appendix A, Figure 17).

The team led by Pierre Bouchard excavated the 16C section of the site (Appendix A, Figure 17), immediately east of 16D. Their objectives were also to excavate down to the original beach level the site was situated on, though it was realised this was not possible due to the time constraints of the field school. The dividing berm from the 11C excavations was excavated, and a smaller area was concentrated on. This area contained the easterly extension of the privy system (16C26), and the foundation supports of a building dating to 1875.

The numbering system used in the excavations was an alphanumeric system. The operation number 16 was assigned to all the 1995 excavations, and each area of the site or sub-operation had a letter designation. As layers were opened, they were numbered sequentially. In the 16D area of the 1999 excavations, for example, the first layer excavated was 16D1, while the final layer was 16D43. Structures such as the Dauphine Battery were normally given numbers over 100, resulting in the Dauphine Battery being denoted 16D100, while the western section of the large privy structure became 16D200. After mechanically removing the sand that was placed over the site to protect it after the 1993 excavation, all subsequent layers were excavated by trowel. However, once it was decided to excavate down through the fill in front of the junction of the de la Chesnaye wharf and the first Dauphine battery, a backhoe was used for the removal of this level, 16D3. Two large berms, running north-south and east-west through the 16C and 16D sub-operations were used to study stratigraphy and to ensure stability.

During the excavation, four main areas of the site, 16C, 16D, 16E, and 16K were sampled for insect and parasite remains. These areas contained two privies, a well or cistern, and a drain; structures significant in the interpretation of the site. Each team leader ensured samples were consistently taken for the insect and parasite studies. For the archaeoentomological analysis, a minimum of two 2-litre bags of soil were taken from each level opened if the layer permitted, while two small 100 ml bags of soil were taken from the same layers for the archaeoparasitological analysis.

The interpretation and reporting of the 1995 excavations was part of Pierre Bouchard's Master's thesis project, and the following selected description and interpretation of the excavations refer to his site report and thesis (Bouchard 1997, 1998).

Sub-operation 16C

The 16C area of the site measured 7 m north-south and 4.8 m east-west. It included the previously excavated areas 11B, 11C and 11D (Bouchard 1997:5). The berm left from the 1993 excavation was excavated and then the 16C team primarily focussed on the excavation of the 16C26 privy and the stone columns which were interpreted as the foundations of a building structure dating to 1875 (Bouchard 1997:36).

These stone columns (16C100, 16C110, 16C120 and 16C130) (Appendix A, Figure 20), made of mortared schist and slate, were excavated in the southern part of the 16C trench. These align with 16D102, 16D105 and 16D106. The eighth column, 11A130, was destroyed in 1993 (Bouchard 1997:5). They were roughly a metre in height and were .5 m to 1 m square. Two of them, 16C100 and 16C130 were dismantled, and the two series of wooden planks found underneath these columns were interpreted as a sill (Bouchard 1997:6). Two of the columns, 16C130 and 16D106 were constructed within the 16D200/16C26 privy. The deposits around these two columns date to 1870 and 1875 respectively, and the building appears on insurance maps around 1875 (Bouchard 1997:36-37).

The western section of the privy system, 16C26, measured 3.4 m east-west and 1.75 m north-south. Its base was constructed of planks running east-west at 3.3 m asl. Vertical planks 20 cm wide made up the north and south walls of the privy (Appendix A, Figure 21) and rose to a height of 1.1 m (Bouchard 1997:6). A drain (11B50/11B55) (Appendix A, Figure 19) at the eastern end of the structure was excavated in 1993, while the 16C130 column interrupted part of the northern side of the structure. At its western end, 16C26 connects to the 16D200 section of the privy complex where its bottom planks overlay those of 16D200 (Bouchard 1997:6).

The layers surrounding 16C26 included 16C1, 16C6, 16C24, 16C25, 16C27, 16C29, 16C34 and 16C40. The fill dated to the construction of the second Dauphine battery, between 1742 and 1745, and included green glazed earthenware, Beauvais and Saintonge ceramics (Bouchard 1997:16). The fill within the privy contained many layers which were excavated arbitrarily in five to ten centimetre layers. These included layers 16C44, 16C45, 1647, 16C50, 16C52, 16C53, 16C55, 16C58, 16C61, 16C63, 16C64, 16C65, 16C67, 16C71, 16C72, 16C73, and 16C75 which lay between the 16C130 column and the eastern end of the privy (Bouchard 1997:11). The soil in these layers was a black clayey loam with some schist. Bouchard concluded that the bones of mammals, birds and fish made up almost 75 percent of the inclusions in these samples. Ceramics made up another six percent of the deposit, and were primarily fragments of ball clay smoking pipes, which were marked 'McDougall/Glasgow', manufactured between 1846-1967. Other ceramics included flown blue, blue transfer-printed ware, Rockingham and willow ware. Glass artifacts included pieces of lead glass goblets and dark green glass which indicated traces of Ricketts moulds dating their manufacture to between 1821 and 1840 (Les Recherches Arkhis 1987:150).

The western section of 16C26 between the 16C130 stone column and the north-south berm of the 1995 excavations was arbitrarily excavated in eastern and western halves with the goal of locating the junction between the 16C26 and the 16D200 sections of the privy system. Layer 16C54 was located east of this arbitrary line and was 26cm thick. It was a black clayey loam layer which had a spongy texture and contained many bones, window glass, bottle glass, and goblet fragments as well as forged and cut nails (Bouchard 1997:13). Similar layers, 16C56, 16C57, 16C60, and 16C59 were excavated in this section of the privy. Beneath these layers was 16C68/16C70, a layer of organic clayey loam which differed from the previous layers as it was darker and gave off a strong odour. It was contained to the west by the north-south berm and to the east by 16C69. The interface between 16C68/16C70 and 16C69 was delimited by a plank laying edgewise on the bottom of the 16C26 structure. The ceramics found in this assemblage were willow and flown blue white wares as well as blue, red and brown transfer-printed ware, and yellow ware. McDougall/Glasgow pipe fragments were also found as well as bone objects, window glass, green glass bottle fragments and bird and mammal bones (Bouchard 1997:15). 16C69 was a compact dark clayey loam and was almost entirely made up of mammal, bird and fish bones (Bouchard 1997:15).

On the very bottom of the privy, between the edgewise planks mentioned above (Appendix A, Figure 22), was the 12 cm thick layer, 16C74, a brown sandy loam. Beauvaisis earthenware, shell-edged pearlware and blue transfer-printed ware was found as well as blue transfer-printed ware marked 'Davenport, 1836'. Green and red transfer-printed ware, Rockingham, ironstone, and yellow wares were also found along with pipe fragments, window glass, a wax seal, as well as the ubiquitous bird, fish and mammal bones (Bouchard 1997:15-16).

Layers related to the building and renovation of the 1875 building in this area include 16C14, 16C32, 16C37 and 16C41. These layers contain traces of building renovations, which disturbed the surrounding soil, and mixed with fill dating to the pre-Conquest period. The construction and renovation activities date between 1875-1900 (Bouchard 1997:38). Layers 16C14 and 16C32 were a dark grey sandy loam, and contained ash, mortar and schist. The artifacts were of a mixed nature as there was Staffordshire fine wares, brown and white faience, creamware, pearlware, stoneware, blue transfer-printed ware, chert, charcoal, oysters, and bones (Bouchard 1997:9). Layer 16C41, a compact clayey loam, contained brick and mortar inclusions (Bouchard 1997:10).

Upper layers between the north-south berm at the western edge of the 16C sub-operation and the 16C130 stone column were 16C16 and 16C18, while east of 16C130 lay 16C28, contemporaneous to the previous two layers. These black compact layers, within the physical limits of 16C26, date to the late 1800's and are under planks of the 1875 building. They contained mortar and brick inclusions, ironstone, window glass, yellow ware, and bird, fish and mammal bones (Bouchard 1997:10). Bouchard states the contents of these late nineteenth-century layers filled in a depression, perhaps not unlike the late nineteenth-century fill on top of the 16D200 privy, as the contents beneath began to compress. The ironstone found in these levels dates them to the use of the 1875 building (Bouchard 1997:40).

The entire 16D200/16C26 privy was constructed around the mid-nineteenth century. In the 16C26 section of the privy, the deposit 16C74, laying directly on the privy bottom contains ceramics that date to around 1850, while the layers above date to 1860-1870 (Bouchard 1997:33). Layers 16C68 and 16C70 are of a similar soil type, and contain artifacts with matches to the 1860s layers of 16D200 (16D27, 16D31, 16D33, 16D35, 16D39, 16D40, 16D41), and the 16D203 drain, (16D30, 16D34, 16D38, 16K5). 16C68 and 16C70, therefore date to the 1860s (Bouchard 1997:34). The many layers including 16C44, 16C45, 1647, 16C50, 16C52, 16C53, 16C55, 16C58, 16C61, 16C63, 16C64, 16C65, 16C67, 16C71, 16C72, 16C73, and 16C75 date to around 1870, as they are above the lowermost deposit on the privy bottom (16C74), and predate the construction of the 1875 building (Bouchard 1997:34).

Sub-operation 16D

The 16D excavations focussed on the excavation of the 16D200 section of the privy, the 16D203 drain, three stone columns (16D102, 16D105, 16D106) and the area in front of the Dauphine Battery with the goal of reaching the original beach level .

As in the 16C sub-operation, it was quickly realised that the goals for the 16D excavation were too ambitious for the time available. While the 16D200 privy structure, the adjacent drain (16D203), and the 16D102, 16D105, and 16D106 were excavated manually, the 1745 fill (16D3) in front of the Dauphine Battery (16D100) and the de la Chesnaye wharf (16D101) was cleared away with a backhoe to reach the historical beach level (Appendix A, Figure23).

Three stone columns, 16D102, 16D105 and 16D106, (Appendix A, Figure 20), noted during the 1993 excavations (Goyette 1994:11-12) were excavated. They were roughly 80 cm square and composed of stones joined with mortar. While 16D102 and 16D105 are situated in the fill in front of the Battery, 16D106 was located within the 16D200 privy (Appendix A, Figure 18). As previously stated, they align with those dug in the 16C area of the site (Bouchard 1997:18), and, have been interpreted as the foundations for a building dating to 1875.

During the excavation with the mechanical excavator in front of the Battery and the de la Chesnaye wharf (16D101), the side of a privy (16D25) was noted in the south wall of the trench flush with the face of the stone wharf (Appendix A, Figure 24). This is the same privy that was located during the 1992 excavations (10A10). This structure, seen only in the wall of the excavation trench, was sampled and not excavated. A piece of a creamware chamber pot was noted in the profile (Bouchard 1997:17-18). Results from the 1992 excavations of this structure indicate it was constructed some time after 1825 (L'Anglais 1998:89), and was abandoned sometime 1850-1875 (Bouchard 1997:31).

The 16D200 privy (Appendix A, Figure 2) measured 2.45 m east-west and 3.15 m north-south and is made of wooden planks, 10 of which run in a north-south direction on the privy bottom. Like the 16C26 privy, 16D200 was 1.1 m deep and has cross planks on its bottom to secure the privy floor. This part of the privy was also served by a drain (16D203) located in the northwest corner of the 16D200 area. As the east-west berm of the 1995 excavation dissected the privy structure into northern and southern sections, they were excavated accordingly. Upper layers in the northern part of the privy (16D4, 16D6, 16D7, 16D10, 16D12, 16D13) (Appendix A, Figure 25), contained lots of mortar, ash, stones, wood, a few ceramics and bones (Bouchard 1997:20-21), and had the appearance of destruction layers. Beneath these levels, the artifacts changed and different soils were noted. These different soils (16D18, 16D20, 16D21, 16D22, 16D23, 16D24, 16D26, 16D29, 16D32) (Appendix A, Figures 25, 26) now varied from sandy to clayey, though these levels also contained bricks, mortar, wood and stone inclusions, and were excavated in 10-cm levels. This section of the privy fill was 1.3 m north-south and 1.4 m east-west (Bouchard 1997:21). These layers contained flown blue, ironstone wheat ware some of which had were marked with 'Furnival & son' (1871-1890), and 'Robert Cochran' (1863-1918), and a date within a diamond-shaped mark of 8 December 1858. There were several pipe fragments with the marks of 'Henderson/Montréal' (1847-1876), 'W & D Bell/Quebec' (1862-1881), 'McDougall/Glasgow' (1846-1967), 'Coghill/Glasgow' (1826) and 'Noël Paris' (+1890). A coin was also found dating to 1884 (Bouchard 1997:22). Bouchard has interpreted the levels listed above, as fill placed into a depression, as the lower layers of the privy decayed and compressed over time (Bouchard 1998:73). Stratigraphically these layers sit between one dated to 1860 and a twentieth century deposit (16D4, 16D6, 16D7, 16D10, 16D12, 16D13) (Bouchard 1997:39), thus Bouchard arrived at a date of between 1880-1900.

Beneath this 1880-1900 deposit were very organic soil layers that gave off a strong odour. These layers, 16D27, 16D31, 16D33, 16D35, 16D39, 16D40, and 16D41, totalled 77 cm of deposit, and descended to the privy bottom. Like the previous levels, they were excavated arbitrarily. Many colours of transfer-printed wares marked with 'Davenport' and 'Middlesboro Pottery' were recovered as well as decorated glazed white wares, transfer-printed ironstone, yellow wares, stonewares, and pipes with the marks of 'Murray/Glasgow' (1830-1861) and 'McDougall/Glasgow' (1846-1967). Window and bottle glass, leaded glass goblets, leather objects, fish and mammal bones and oysters were also recovered from this rich deposit. Bouchard dates these levels to around 1860 (Bouchard 1997:22).

The two previous sections of the privy described above were intact, while the areas to the south and east, had already been at least partially excavated in 1993 (Appendix A, Figure 27). On the eastern side of the north-south berm, levels 11C29, 11C31 and 11C33 (Appendix A, Figure 28) were excavated in 1993 and were contemporaneous with 16D27, 16D31, 16D33, 16D35, 16D39, 16D40, and 16D41. Insects studied from these layers (Bain 1997a) are integrated into the 1995 study. Level 11C29 was a sandy clay layer which contained bricks, schist, wood and charcoal and a coin dating to 1871 (Goyette 1994:32). The layer underneath, 11C31/11C33, was darker and appeared to be more organic, clayey and compact. It contained strips of leather and fabric, shoe laces, several bottles, forged nails, a lead seal, eggshells, bones, oysters, coffee beans, peach seeds, melon and pumpkin seeds. There were flown blue ceramics some of which were marked 'Daly's Hotel, Montréal' (Appendix A, Figure 29), along with pearlware, salt-glazed stoneware, white wares, yellow ware and pipes fragments bearing the name of 'Henderson/Montréal (Goyette 1994:33-34).

South of the east-west berm, within the privy structure was the stone column 16D106. Several layers on top of the column had been excavated in 1993, while the layers adjacent and east of the column (16D14, 16D15, 16D16, 16D17, and 16D28) (Appendix A, Figure 30), were excavated in 1995. Layers 16D14, 16D15, and 16D16 were made up of a sandy loam with inclusions of brick, stones, mortar and wood. In these layers were blue and brown transfer-printed wares, and white wheat ware as well as pipe fragments marked 'W&D Bell/Quebec' and an 1886 coin. Under this deposit were layers 16D17 and 16D28 which were rich in highly patterned tablewares (Bouchard 1997:23).

A vertical plank separated these levels from those directly north (situated in the north-south berm) which were 16D36, 16D37, 16D42 and 16D43. It appeared to be a separate deposit, but contained artifacts similar to those in the 16D17 and 16D28 layers and included a coin dating to 1876, cut nails and butchered bones (Bouchard 1997:23). Both of these series of levels, 16D14, 16D15, 16D16, 16D17, 16D28 and 16D36, 16D37, 16D42 and 16D43, appear to have been deposited rapidly, due to the almost total lack of soil in and around the artifacts. Furthermore, many of the objects were thrown away whole and there were many examples of some of them. The fill dates to around 1875 indicated by the fact that one of the stone columns (16D106) from the 1875 building sits on top of it and by dating the white ware and ironstone (Bouchard 1998:70).

Layers within the drain 16D203/16K300 (Appendix A, Figure 31) that ran into the northwest corner of the privy included 16D30, 16D34, and 16D38. The drain was 5.25 m in length, running east-west towards the Hunt house and was 70 cm wide. To construct the drain, part of the surface of the Dauphine Battery had to be destroyed (Bouchard 1997:18) (Appendix A, Figure 32). The organic and sandy soil within this structure contained inclusions of wood, stone and mortar, and dated to the same time period as artifacts in layers 16D27, 16D31, 16D35, 16D39, 16D40 and 16D41, confirmed by mending matches between artifacts from the two deposits (Bouchard 1997:24). As the drain would have been blocked by the deposit of these artifacts, Bouchard has interpreted this event, in both the drain and privy bottom, as a discontinuation of the use of the privy around 1860. Some of the objects discarded were almost complete, further indicating they were not longer needed or desired, and were rapidly disposed of (Bouchard 1998:68).

Sub-operations 16K and 16M

Sub-operation 16K was a 2 m square west of the Dauphine battery (16D100) at the location of the 16D203 drain. It was opened to test whether a wall noted in 16G, 16G500, was extended into this area, and to study further the damage to the battery from the drain's construction. The wall under question, 16G500/16K200 did extend into this area and was just south of the drain. The layer 16K5 within the drain (16D203/16K300) (Appendix A, Figure 33), was similar in soil type and inclusions to the drain layers described above and has been dated by Bouchard to around 1860. There were many objects in this layer, which included salt-glazed stone ware, pipe fragments, common earthenware from northeastern England, as well as blue, red, yellow and white-glazed tablewares. Window glass, green bottle glass, clear leaded glass goblet fragments, leather, mammal and fish bones and oyster shells were also recovered (Bouchard 1997:26).

Sub-operation 16M tested whether the drain continued from sub-operation K further westwards to connect with the Hunt house. An area was opened 4.6 m west of sub-operation K and measured 1.2 m by 1.5 m. Two layers were excavated, neither of which indicated an extension of the drain into this area. The alternative hypothesis is that the drain was extended, and its more western traces were destroyed when the drain was no longer in use (Bouchard 1997:27).

Sub-operations 16E and 16F

The goal of sub-operation 16E, a 2 m east-west and 4.5 m north-south excavation along rue Saint-Antoine, was the excavation of 16E100. This 1 m wide stone circular structure (Appendix A, Figure 34) was interpreted as a well when it was discovered during the 1994 excavations of the 1824 house foundations (Dubé 1995:54-55).

Five layers were excavated within 16E100 down to a depth of 1.37 m, the first layer, 16E1, was a brown compact sandy and clayey soil and contained bricks, mortar, schist and wood cuttings. It was excavated to a depth of 19 cm and contained a pipe fragment marked 'TD', Iberian and English earthenware, a green glass vial, a coin dating to 1812, white faience, willow and shell-edged white ware, and a bottle base with a pontil scar (Bouchard 1997:47).

The following layer, 16E2, 13.4 cm in depth, was of a similar soil type, but had more of a grey colour. Like 16E1, it contained bricks fragments, schist, mortar, charcoal and wood fragments. Creamware, pearlware, a clear medicinal bottle, yellow ware, faience, window glass, a silver ladle, bones and a coin dating to 1820 were recovered (Bouchard 1997:47).

Layer 16E3 was a brown sandy clay mixture with inclusions of brick, mortar, stones, and pieces of wooden planks. This 12 cm thick layer contained Iberian earthenware, pearl ware, green bottle glass fragments and a pipe fragment that dated to 1826 (Bouchard 1997:47). 16E4 was a damper layer and was the same colour, with similar inclusions as the previous layer. Some shell-edge pearl ware, Iberian earthenware, blue transfer-printed ware, a stone ware shoe polish bottle, green bottle glass, a leather sole, bones and eggshells were all recorded in this 11 cm level (Bouchard 1997:47-48).

The level beneath, 16E5, was a clayey brown soil of the same colour and was 13 cm think and very damp. Wood chips and the end of a wooden beam were recovered. The beam was found in the centre of the structure and was 23 cm by 26 cm and 50 cm long. Cultural remains consisted of stoneware, green bottle glass, pearlware, metal, nails, shells, and bones. The bottom layer of the structure, 16E6, was a thin layer of mortar with red brick and stone inclusions and contained only a few pieces of green glass and some bones (Bouchard 1997:48). When entirely excavated, there was always between 12-60 cm of water in the structure (Appendix A, Figure 35), though Bouchard was not convinced this was in fact used as a well, contrary to Dubé's interpretations. Bouchard instead suggests it may have been used as a cistern (Bouchard 1997:57). It may have indeed functioned as a well, as one was built by the city on rue Saint-Antoine around 1830, suggesting this area was a viable location for one (Simoneau and Rouleau 1988:85). There was also reasonable water access at the end of rue Saint-Antoine at this time, where the city had constructed public pumps (Simoneau and Rouleau 1988:85). This fact would suggest that it might have been more prudent for that household to have a cistern. The similarity in the deposits within this feature suggests they were filled quickly around 1850. This structure was used for a maximum of 25 years (Bouchard 1997:57), at the end of which it was filled with unwanted trash.

Sub-operation 16F (Appendix A, Figure 34) was to locate the remnants of a latrine that may have been south of the 1824 house foundations, indicated by 1910 insurance maps. Two structures, 16F100 and 16F200 were noted (Bouchard 1997:50), though there was no evidence of latrines found.

Summary of the depositional sequences within the 16D200/16C26 privy

In his analysis of the material culture from the 1995 excavations of the Îlot Hunt site, Bouchard was able to identify five depositional episodes into the 16D200/16C26 privy complex. The first, in the oblong section of the privy, dating to 1850, is layer 16C74, likely missed during a later cleaning episode (Bouchard 1997:33).

The second deposit, around 1860 appears to be the result of a rapid filling of the 16D200 privy bottom (16D27, 16D31, 16D33, 16D35, 16D39, 16D40, 16D41), part of the western end of 16C26 (16C68, 16C70), and the 16D203 drain (16D30, 16D34, 16D38, 16K5). As some of the objects are complete, and the large pieces in the drain would have effectively barred its functioning, this has been interpreted as an abandonment episode (Bouchard 1997:34). The third event is represented by many layers of the oblong section (16C26) represented by many layers (16C44, 16C45, 1647, 16C50, 16C52, 16C53, 16C55, 16C58, 16C61, 16C63, 16C64, 16C65, 16C67, 16C71, 16C72, 16C73, 16C75), which date to around 1870 (Bouchard 1997:34-36).

Another rapid filling episode took place around 1875 (16D14, 16D15, 16D16, 16D17, 16D28, 16D36, 16D37, 16D42, 16D43). These deposits are located in the southern section of the 16D200 area and represents the reuse of the 16D200 section that was previously abandoned, indicating that it had been cleaned out at least once. There is no evidence, adjacent to these levels, of the 1860s depositional event, suggesting it had been cleared away (Bouchard 1997:36)

The fifth and final deposit (16D18, 16D20, 16D21, 16D22, 16D23, 16D24, 16D26, 16D29, 16D32) was between 1880-1990, after the privy had been abandoned. These levels are located roughly in the centre of the privy, and likely filled a depression due to the compression of the organic material in the lower privy levels (Bouchard 1997:38-39). Layers 16C16, 16C18, and 16C28 may also be included as they date to the late nineteenth century and appeared to fulfil a similar function, that is, they filled in a depression, perhaps also due to the compression of the 16C26 levels which lay underneath.

Surveys 16A, 16B, 16G, and 16J

The archaeological surveys 16A, 16B and 16G (Appendix A, Figure 17) were trenches running around the Hunt house, which underwent renovations in 1995. Survey 16A took place along rue Saint-Antoine, and involved the excavation of a concrete pipe installed by Hydro-Québec and of part of the return of the Dauphine battery. The 16B trench was east of the Hunt house, and joined the 16A trench at its northern end. This 10 m stretch behind the east wall of the house confirmed known structures and events, while the 16G trench, near the southeast corner of the house uncovered evidence of a stone building dating to 1815 (Bouchard 1997: 61,73). Area J was a 32.4 m by .5 m trench that ran between the Auberge Saint-Antoine and the Hunt house, intersecting previously studied structures such as part the first Dauphine Battery and cellar of the 1824 house on rue Saint-Antoine (Bouchard 1997:65-67).

3.3 SPECIALIST ANALYSES RELEVANT TO THE 1995 EXCAVATIONS

Four specialised analyses have been undertaken on the 16D200/16C26 privy. Faunal remains were studied from the 1995 excavations (Boucher 1999), while Bouchard (1998) studied the material culture of the privy complex from a socio-economic perspective. A palaeobotanical analysis by Fortin (1996) studied seeds recovered from the 1993 excavation of the privy, while an archaeoentomological analysis (Bain 1997a) investigated similar levels. The results of the archaeoentomological study based on 1993 excavation (Bain 1997a), have been integrated with the archaeoentomological analysis of the 1995 excavation and are presented in chapters five and six. The other three analyses will be discussed briefly, as they are relevant to this project. They are also integrated with the discussion of the insect and parasite faunas in chapter six.

Socio-economic analysis

Pierre Bouchard was responsible for the interpretation of the 1995 excavations and the material culture which it produced, subjects treated in the previous section. To further investigate the rich assemblage of artifacts from the 16D200/16C26 privy, according to his assigned depositional events,

Bouchard used a relative value index on fine white English ceramics according to their decoration (Bouchard 1998:82-83) This 'Miller index' creates a value to which an index is given. The average of these indexes for a certain event or context, may be compared and contrasted against others for interpretive purposes (Bouchard 1998:83). In his initial analyses of the ceramics, Bouchard found a progressively higher index for the depositional episodes dating to 1850, 1860 and 1870, while the 1875 and late nineteenth-century events indicate a decreasing index. This parallels the economic situation of Québec in the last quarter of the nineteenth century as it suffered through a recession (Bouchard 1998:84).

To test these initial findings, Bouchard then compared his results with those from a contemporaneous site, La Grande Place in Québec City's Saint-Roch ward. For this second analysis, the ceramics were regrouped into ceramics of low, medium and high value, giving more of an indication of the socio-economic level of the site's users. The results of this second interpretation are interesting in that they demonstrate the site's inhabitants had more ceramics of high value around 1850 and from 1875 to 1900 (Bouchard 1998:87). When compared to the Grand Place site, the Îlot Hunt levels have higher indexes. While Bouchard feels the two sites have fairly similar indexes around mid-century when the actual objects and site inhabitants are considered (Bouchard 1998:91), the Grand Place inhabitants become progressively poorer as the century progresses, while the inhabitants of Îlot Hunt remained at a higher socio-economic level (Bouchard 1998:91). Bouchard's findings are pertinent to our analysis as they provide valuable socio-economic information about the site's uses that may be later integrated into our discussion.

Archaeozoological analysis

Guylaine Boucher, a Master's student at Université Laval, studied the faunal remains recovered from the 1995 excavations of the privy (16D200/16C26). To organise her data, she followed Bouchard's chronological interpretation of the privy structure, which outlined five distinct depositional events between 1850-1900 (Bouchard 1998:66-70, 73-74). His hypothesis of an improved socio-economic standing for the site's inhabitants post-1870 was also taken into consideration in Boucher's analysis.

Boucher found that if studying the sheer diversity of species, as an indicator of the purchasing power of the site's occupants, they did have in fact good economic standing throughout the fifty-year period (Boucher 1999:101). Other ways of elucidating more sensitive material from the data are to study the age of mammal individuals at death and to look at butchery cuts (Boucher 1999:102).

By studying mammal age at time of butchery, Boucher was able to confirm her findings, which indicated that the inhabitants of Îlot Hunt consumed many young animals, some of which were probably raised on site. This interpretation is drawn from the butchering wastes and different animal body parts found. If representatives of all the animal body parts are found, then they were likely raised on the site (Boucher 1999:102). The slight decrease in this presence coincided with the economic downturn of the 1870's, followed by a strong rise in the consumption of veal, lamb and young pigs around 1875 and between 1880 and 1900 (Boucher 1999:102).

Her third area of study was of butchery cuts found on sheep, beef and pork bones. First she identified mutton as the most commonly consumed animal at the site, and then the proportion of cheap versus costly meat cuts was compared. For beef, the circa 1875 context saw a sharp rise in the use of inexpensive cuts of beef, again indicating tougher economic conditions, though better cuts were again favoured in the late nineteenth century. Sheep, except for the deposits dating to around 1850, was usually represented by good cuts of meat (Boucher 1999:103). In the first deposit dating to 1850, there were very few pork bones, but they were followed by a dominance of better cuts around 1860, which levelled out around 1870. An increase in quality pork cuts was seen in the fourth and fifth phases (Boucher 1999:104). Consumption habits at the site did, in fact, change over this fifty-year period Boucher concludes, but were also the result of personal choice and ethnic preferences, which may account for the large amount of sheep consumed. Mutton consumption declined during the 1860s coinciding with the reduced English population on the site, leaving a predominantly Irish population on the site until the turn of the century (Bouchard 1998:32).

As seen with Bouchard's material cultural study, Boucher's findings provide relevant supplementary data about the site's users and their consumption patterns. The insect remains recovered from the Îlot Hunt samples should also indicate the presence of animals, and their subsequent butchery in the yard area. By having this archaeozoological information available, it provides a mechanism by which we may assess the accuracy of the insect results in reflecting their depositional contexts.

Palaeobotanical analysis

The palaeobotanical material studied from the 1993 excavations (Fortin 1996) pertinent to this project was sampled from five layers including 11C29 and 11C33 (Appendix A, Figure 28), in the northeastern section of the 16D200/16C26 privy. Layers 11B46 and 11B51 were part of the contents of the oblong extension of the privy (11B47/16C26), while 11B53 was sampled from the drain (11B50/55) (Appendix A, Figure 19) at its easternmost end.

In level 11C29, 67 of the 73 seeds recovered were raspberries. The others included strawberries, field sorrel, sedge and campion seeds. The remains in this level were all small seeds from edible plants and herbs, perhaps the result of defecation, cleaning of a chamber pot or contamination (Fortin 1996:11-12). The 11C33 layer contained a lot of organic material, including evidence of pumpkin, melon, peach and coffee, while raspberries, figs and strawberries dominated the assemblage. Other berries identified were blueberries, cherries, and elderberries. According to Fortin this is a typical latrine assemblage (Fortin 1996:11-12).

While it was quite diversified, this level was not particularly rich in plant remains, and it has been interpreted as a mixed deposit of faecal remains and food wastes. The amount of weeds found was slightly more than normal, and suggests the presence of animal bedding (Fortin 1996:13).

11B46 and 11B51, dating from 1860-1875, were arbitrary levels at the eastern end of 16C26. The samples contained the four edible species of raspberries, cherries, figs and grapes, while the weed species, 36 percent of the assemblage, represented a natural environment. Fortin concludes they do not appear to be from privy deposits (Fortin 1996:10).

A drain located just east of the 16C26 privy section (11B50/11B55), contained the sample 11B53, and is contemporaneous with the previous sample. Only five percent of the samples were made up of 'natural species', while there were four edible species (Fortin 1996:10) According to Fortin, these samples do not look like latrine fill, and the few traces of the natural environment suggest the privy and drain were covered (Fortin 1996:10). After comparison with a similar drain excavated at Place Royale, it appears that the Hunt drain may have been removing rainwater and wastes that were contaminated by the faecal material of the pit. Fortin concluded that these 11B soils, like many urban archaeological deposits, had mere traces of faecal matter (Fortin 1996:10).

The palaeobotanical information derived from Fortin's analysis indicates dietary information, and perhaps more importantly, contextual information about the levels themselves. Just as in the archaeozoological analysis, Fortin's finding will be integrated with the outcome of the insect and parasite remains. Her results, indicating drain contents, a 'typical' latrine assemblage and the possible presence of animal bedding are pertinent to our discussion of the 16D200/16C26 privy, and provide a comparable environmental data set to the insect and parasite remains.

Summary of Îlot Hunt levels from 1993 and 1995 excavations

To better understand the many contexts described in the preceding sections, the following table (Table 3.1) charts the levels mentioned from the 1993 and 1995 excavations, as well as the specialist analyses, in association with their historical and structural settings. Levels in bolded text were processed for insect and/or parasite remains.

Table 3.1 Summary of levels from 1993 and 1995 excavations

Date	Context	Corresponding Levels				
c. 1742-1745	Fill related to the construction of the Second Dauphine Battery	**16C1**	16C6	**16C10**	**16C24**	16C25
		16C27	**16C29**	16C34	16C40	16D3
c. 1850	Drain running diagonally across yard	11C11				
c. 1850	Privy 16D25/10A10	**16D25**				
c. 1850	Well/cistern 16E100	**16E1**	**16E2**	**16E3**	**16E4**	**16E5**
		16E6				
c. 1850	Privy section 16C26	**16C74**				
c. 1850	Drain at eastern end of 16C26 privy section	11B46	11B51			
c. 1860	Privy section 16C26	16C68	**16C70**			
c. 1860	Privy section 16D200	**11C29**	**11C31**	**11C33**	**16D27**	**16D31**
		16D33	**16D35**	16D39	**16D40**	**16D41**
c. 1860	16D203/16K300 drain	**16D30**	16D34	**16D38**	**16K5**	
c. 1870	Privy section 16C26	**16C44**	16C45	**16C47**	16C50	**16C52**
		16C53	**16C54**	**16C55**	**16C56**	**16C57**
		16C58	16C59	16C60	**16C61**	16C63
		16C64	**16C65**	**16C66**	16C67	**16C69**
		16C71	**16C72**	16C73	**16C75**	
c. 1875	Privy section 16D200	16D14	16D15	**16D16**	**16D17**	**16D28**
		16D36	16D37	**16D42**	16D43	
c. 1880-1990	Layers related to the renovation of the c. 1875 structure	**16C14**	16C32	16C37	**16C41**	
c. 1880-1990	Layers located in a depression under the c. 1875 structure	**16C16**	**16C18**	**16C28**		
c. 1880-1900	Layers infilling a depression over the 16D200 privy section	**16D18**	16D20	16D21	16D22	**16D23**
		16D24	**16D26**	**16D29**	16D32	
20th century	Destruction layers	**16D1**	**16D2**	16D4	**16D5**	16D6
		16D7	16D10	16D12	16D13	

3.4 MEDICAL AND SANITARY PRACTICES AT ÎLOT HUNT

Medical and sanitary practices are represented at the site by material culture in the form of artifacts and structures related to sanitary practices at the site. Documentary evidence such as infrastructure and demographic data, provide valuable information as to the services available to the site's inhabitants, while socio-economic information may be gained from the demographic information.

Material culture and structural remains dating to c. 1850

On the 1850 Îlot Hunt site, there were numerous structures related to sanitary practices which included two privies, a well or cistern that was filled with trash and a drain (11C11) running diagonally across the yard to the southeast (Appendix A, Figure 36).

The 16D25 privy was in use at this time as was the large 16D200/16C26 privy structure, indicated by the 16C74 deposit in the 16C26 section. As this two-part privy was built around the same time, the 16D200 section was also in use. The well or cistern structure, 16E100, was used at this time, to deposit trash. While the 16D203 drain connected to the 16D200 section, was filled around 1860, it was likely constructed and used during the 1850s, as was the drain at the south-eastern section of the structure, 11B50/11B55. The drain which lay diagonally across the yard, 11C11, functioned at this time though it did not appear to properly connect to the 16D200 section of the privy (Goyette 1999:141), and may have been used primarily to serve the rue Saint-Antoine houses. It is unclear whether the 16D203 drain did in fact connect to the Hunt house. It may have also been connected to the 1815 building behind the Hunt house (Appendix A, Figure 18) or to another secondary building, which appears on 1875 plans, represented by the columns found during the 1995 excavation (Appendix A, Figure 20). The Hunt house likely benefited from a drain on rue Saint-Pierre. These drains are mentioned in a mid-century report from Québec City's Health Office (Bureau de santé, canalisation de la ville 1849:14). As the municipal sewer and water services are just being constructed around 1850, they would not have been available to serve the houses on rue Saint-Antoine. There was, however, a drain running along the southern side of rue Saint-Antoine at least as late as 1838, and documentary evidence indicates that the owner James Hunt had to repair it. Evidence for the structure was found by an archaeological investigation (Simoneau and Rouleau 1988:103-107), though it is uncertain whether this structure was still in use as late as 1850.

The few material culture remains indicating sanitary or medical practices from this period (see Table 3.2) include a chamber pot (16D25), a glass vial, a medicinal bottle, and a bottle which contained shoe blackening or polish (16E100), and a pitcher from the 16C74 deposit of 16C26.

Table 3.2 Material culture related to hygienic and sanitary practices recovered from the 1995 Îlot Hunt excavations

Object	c. 1850			c. 1860			c. 1870	c. 1875	1880-1900	Total
	16D25	16C26	16E100	16D200	Drain	16C70	16C26	16D200	16D200	
Ampoule				1						1
Brush				1						1
Chamber pot	1				5		1	3	1	11
Comb					1				3	4
Glass vial			1	3	1	1	2	2		10
Hair brush				1			1			2
Hair dye bottle					1					1
Medicinal bottle			1	2			1	2	6	12
Ointment jar					1			1	1	3
Perfume bottle								10		10
Pitcher		1			1			4	3	9
Shoe polish			1				2	5		8
Toothbrush							1		1	2
Total	**1**	**1**	**3**	**8**	**10**	**1**	**8**	**27**	**15**	**74**

Material culture and structural remains dating between 1860-1875

For this fifteen-year period, the large 16D200/16C26 privy structure and the 16D203 drain as well as continued use of the 11C11 drain (Goyette 1999:141) represent the sanitary structures in use (Appendix A, Figure 37). At this time, the 16D203 drain was not used as a drain, but instead served as a trash receptacle, while the 11C11 drain was still in use for part of this time period. The majority of the 16C26 structure was filled around 1870, while two distinct depositional episodes, dating to 1860 and 1875 were identified for 16D200. The 1875 deposit has been identified as the final use of the privy. The material culture from this fifteen-year period is represented by eight chamber pots and five pitchers (Appendix A, Figure 38). Personal grooming is evidenced by two hairbrushes (Appendix A, Figures 39, 40), a toothbrush (Appendix A, Figure 39), a comb, and another brush handle, a bottle of hair dye, 10 perfume bottles and seven bottles of shoe blackening or polish. Medicinal practices are represented by an ampoule, five medicinal bottles (Appendix A, Figure 41), two ointment jars, and nine glass vials (Appendix A, Figures 43, 44).

Material culture and structural remains dating from 1880-1900

By the late 1800s there are almost no sanitary structures in use on the site itself as the large privy structure had been filled, though a sewer pipe, dating to the late nineteenth century was found in the 1825 rue Saint-Antoine house foundations (Dubé 1995:47,56). More evidence of the installation of sewer service was found on rue Saint-Antoine in the form of a 1.15 m by 1.2 m red brick sewer collector (CeEt-110-2A6), running east-west along the street, and was accompanied by the lines for the aqueduct and an old gas conductor (Simoneau and Rouleau 1988:102). It was likely installed in the late 1800s or early 1900s (Simoneau and Rouleau 1988:89). In their investigations, Simoneau and Rouleau found evidence of the connecting of the Vallerand building (built by Chillas in 1822) to the rue Saint-Antoine sewer pipe.

Material culture remains from late nineteenth-century deposit on the location of the abandoned privy include a chamber pot, and three pitchers, a toothbrush (Appendix A, Figure 39), three combs, six medicinal bottles and an ointment jar.

Documentary evidence related to sanitation at Îlot Hunt

Documentary evidence related to health and sanitary practices represents two main areas of information. Demographic data from the site must be considered if an understanding of the inhabitants' health is to be interpreted via the preserved parasite remains. While it is known, through tracing home addresses in the city directories, that some of the merchants did not live on the site, by working there, they would have contributed to the refuse, as well as the flora and fauna. The fifty-year period between 1850-1900 was one of great activity on the Hunt site, and by combining census data, city directories, and evaluation roles, we have a fairly good idea of who was working and living on the Îlot Hunt site. The second source of data concerns documents related to municipal sanitary infrastructures that would have affected the users of the property.

Demographic data

Mackay's Quebec Directory for 1848-9 lists three commission merchants and agents for Hunt's Wharf, a general merchant and two general brokers (Mackay 1848). Evaluation roles indicate that a sailmaker also worked in Hunt's wharf while the houses on rue Saint-Antoine contained a tavern and another sailmaker (AVQ, Séries ressources financières, Rôles d'évaluation et d'imposition, 1848). It is interesting to note that evaluation roles for the following year list a temperance house instead of a tavern on rue Saint-Antoine as well as a sailmaker and a labourer (AVQ, Rôles d'évaluation, 1849). Census data combined with the evaluation roles for 1851 total 38 people at the site (Bureau d'enregistrement et de statistiques 1853-1856).

Seven years later, in 1858, the merchants are listed as wine and flour merchants, a commission merchant and a general merchant, and an office of the Phoenix Fire Insurance Company (Cherrier and Hamelin 1858). The 1861 Census of Canada for the Saint-Pierre ward indicates 20 people on the site (Census Department 1863-1864).

In the mid-1860's Hunt's wharf lists two flour and provision merchants, three insurance agents, and the United States Consul. The Hunt house at 66 Saint-Pierre is listed as containing the hardware store of Michon and Talbot, with Etienne Michon listed as a resident (Cherrier 1864). The 1863 evaluation roles also indicate a jeweller and another merchant and store in the Hunt house (AVQ, Rôles d'évaluation, 1863). Other businesses nearby on rue Saint-Pierre at this time were law offices, leather workers, hotels, the Belgium Consul, fruit sellers, tobacconists, commission merchants, ship chandlers and general grocers (Cherrier 1864). Evaluation roles from 1863 show a tavern, a sailmaker and three merchants in the rue Saint-Antoine buildings (AVQ, Rôles d'évaluation, 1863).

Cherrier's Québec Directory for 1870-1871 lists the occupants of Hunt's Wharf as merchants, insurance agents, wine and fish merchants, and a coal oil and lamp grocer as well as the consulate for the North German Confederation (Cherrier 1870). Evaluation roles indicate a jeweller, a broker and a wholesale merchant in the Hunt house while a stevedore, messenger and tavern keeper are shown as proprietors or tenants on rue Saint-Antoine (AVQ, Rôles d'évaluation, 1870). In total, 33 people were listed on the premises during the census (Department of Agriculture 1873-1878).

By 1877, the city addresses match those used today. The rue Saint-Antoine houses still list a saloon or tavern, a boatman and a commission agent. The Hunt house contained an insurance agent, ships brokers, and a ship chandler (Cherrier 1877).

The 1881 Census shows 33 people at Îlot Hunt, almost unchanged since the last census (Department of Agriculture 1882-1885). The houses on rue Saint-Antoine contain a saloon and a cabinetmaker. The Hunt house has an insurance agent, ship brokers, a notary and a ship chandler along with a Mrs. Martin O'Flaherty, presumably a lodger at 68 rue Saint-Pierre (Cherrier 1880). Since 1883, the rue Saint-Antoine store occupied by Hunt Brock & Co., is now shared with A. E. Vallerand a seller of tablewares (Leclerc 1998:28-29).

As of 1900, the Marcotte Annual for the city lists two of the rue Saint-Antoine houses as residences with 8-10 still functioning as a restaurant, and the occupant of 14-16, a store keeper. The former wharf is still occupied by A. E. Vallerand. The Hunt house, at 66 rue Saint-Pierre, contains Royal and Queen's Insurance and the vice-consul for Montevideo, while 68 housed a lumber store and a Miss K. O'Flaherty. Number 68 ½ Saint-Pierre was a wine seller's, while next door was a barbershop (Marcotte 1900). The Census of Canada lists 28 people at the site (Agriculture Canada 1902-1906).

Documented infrastructure data

Documentary evidence indicates that there were drains on rue Saint Pierre and rue Saint-Antoine throughout the nineteenth century, though these sources are not always clear as to when such structures actually existed.

In the 1830s, municipal authorities extended rue Saint-Antoine with a wharf that was to reach the low tide level (Appendix A, Figure 14). The street also had a well and public pumps at the end of the wharf (AVQ, Procès-verbaux des Juges de Paix, Résolution 757, 4 juin 1838, cited in Simoneau and Rouleau 1988:85). Documentary sources also indicate there was a drain on rue Saint-Antoine as early as 1830. That year, the drain had been damaged by ice, and the roads inspector was to investigate and do the necessary repairs (Procès-verbaux des Juges de Paix, Resolution 1033, p. 433, 14 mai 1830 cited in Simoneau and Rouleau 1988:105). As previously stated, James Hunt was to pay for further repairs in 1838 (Simoneau and Rouleau 1988:106). The drain may have existed up to and beyond 1850, the beginning of the period which we are concerned with.

Another source indicates that there was drainage available via the rue Saint-Antoine wharf. In his 1852 report, the road inspector stated that sewer systems installed in Lower Town's wharves should have large sewers conduits so that they did not have to be cleaned manually, which was the current practice. Streets traversing wharves, carrying debris to the river were to have 56 by 36 inch sewer conduits (Rapport de l'inspecteur des chemins 1852:13). De la Montagne was to drain down to the river along rue des Soeurs (today Côte de la Montagne), and was to have a second class drain of 36 by 24 inches. Rue Saint-Pierre is not mentioned in the report (Rapport de l'inspecteur des chemins 1852:14).

It appears that rue Saint-Pierre was macadamised or paved with stones in 1827 (Laframboise 1991:356). The city's Health Office published a report in 1849 stating some of the city streets were not dry for even six months of the year, and at least a part of the existing canals were to be deepened. Rue Saint-Pierre was to receive 400 feet, though the source is unclear whether 400 new feet of drains were to be installed or deepened (Bureau de santé, canalisation de la ville 1849:14).

Îlot Hunt 1850-1900: a textual and material view of sanitation and health

The fifty-year period between 1850 and 1900 is reflected on the site in five depositional episodes found in two privies, a drain, and a well or cistern structure. These events may be tied into information about the site's inhabitants, available documentary sources, and material culture, and thus be considered a reflection of the changing nature of health and sanitation in nineteenth century Québec City (see also Goyette 1999:160-165).

During 1849, a municipal board of health was formed as the city was suffering from a particularly bad cholera epidemic (Bilson 1980:114). This situation may have provided the impetus to clean the adjacent dwellings as well as the yard area, resulting in the circa 1850 deposits excavated in 1995. As this epidemic pre-dated the installation of sewers and piped water in the city, drains were still used to remove wastes. Such features existed since at least 1830 on rue Saint-Antoine, and Saint-Pierre, and were likely still in use considering the 1849 report on the channels in the city (Bureau de santé, canalisation de la ville 1849:14), and the 1838 fixing of the rue Saint-Antoine drain. The nearby river may have been the water supply for the site as public access to the river was ensured at the end of rue Saint-Antoine.

The Îlot Hunt site had a sanitary system in place, needed by the many structures and businesses using the site at this time. As there was no city infrastructure available to remove wastes, the focus would have been the structures available on the site. Around 1850, a former well or cistern, and a small privy (16D25) were filled with trash, perhaps in an effort to render the area cleaner and remove the much feared source of putrefaction represented by garbage. A large privy, 16D200/16C26 was used for the first time (16C74) and it likely received liquid wastes such as domestic water through an incoming drain (16D203), while an outflowing drain, 11B50/11B55, was located at the other end of the structure and may have initially flown out to the river. The 11C11 drain may have served the house on rue Saint-Antoine dating to 1824, as the restaurant or tavern may have needed more drainage than was available along the streets if they had latrines available to their patrons. The few artifacts related to hygienic and medical practices, indicate the people using the space were not particularly wealthy, as they were only a single, bottle, vial, and chamber pot, not an assemblage one would expect from middle class households. The site supported a population of 38 people at this time, of varied socio-economic standing though a substantial percentage were of a lower socio-economic standing including unskilled labourers (Bouchard 1998:32).

In 1860, the drain behind the 1824 house is still in use, as is the large privy system. The privy was cleaned at least once, missing the circa 1850 deposit, and was now in its second use phase. This second event is evidenced by the depositing of material into the 16D203 drain rendering and into the 16D200 privy bottom. If this drain did, in fact connect to the Hunt house, perhaps it is out of use as rue Saint-Pierre now benefits from city water and sewer services. As this street contained many of the city's financial institutions amongst other important businesses owned by merchants and other entrepreneurs, it would not be surprising that they would be one of the first areas of the city able to connect to such services. Furthermore, the 1849 report on the rue Saint-Pierre stating it was to receive better services (Bureau de santé, canalisation de la ville 1849), may have encouraged infrastructure installation.

Around this time, the site contained several merchants, up to three insurance agents, stores, and the United States Consul. The census taken in 1861 indicates 20 people on the site and there was a reduction in the number of unskilled labourers living here (Bouchard 1998:32). Deposits from the 16D200 privy base and the 16D203 drain indicate a richer assemblage of artifacts than those recovered from the previous decade. Several chamber pots, medical bottles, two brushes, a comb, and a hair dye bottle along with vials and ampoules indicate more attention to personal sanitation and hygiene than the 1850 assemblage indicated, and the levels dating to the 1860 deposits indicates some care in clearing the area.

The 1860s were important years in Québec City for the improvement of sanitary infrastructures. By 1863 the city had almost 100,000 feet of sewer lines (Drolet 1967:269) and water service to 3266 buildings (Hare et al. 1987:296). This availability was perhaps the impetus to discontinue the use of the 11C11 drain and to fill the 16C26 section of the privy by around 1870, if such city services were available. Due to the fear of the coming cholera epidemic, a sanitary by-law was passed in 1866 (Dagneau et al. 1983:156). Cholera would strike again in 1871, and for the last time in 1873 (Bilson 1980:3). The rapid infilling of the 16C26 section of the privy may have been in part, due to the fear of this dreaded disease. While some city services were available, the mechanism for disease transmission was not yet clearly understood, and the eliminating of wastes from the yard and local buildings would have assuaged fears of the disease.

The 1871 census data indicates an increase in the site's inhabitants to 33 (Department of Agriculture 1873-1878). These inhabitants continue to represent local businesses as well as a few unskilled labourers (Bouchard 1988:32). The artifacts found, which include a hair brush, a toothbrush, bottles of shoe blackening, a medical bottle and a couple of vials would appear to represent the varied socio-economic standing of the site's population.

Around 1875, the large latrine continues to be the only sanitary structure in use on the site. The deposit from this time period is from the southwest corner of the 16D200 section of the privy, and it appears that this part of the privy was cleaned out and refilled previous to the time of this, the final privy use. The 1875 deposit is made up of chamber pots, medical bottles, vials, an ointment jar, as well as several bottles of shoe blackening and perfume bottles, non-essential items indicating a degree of disposable income, previously unavailable to users of the site. If we consider this deposit relative to the changing socio-economic portrait of the site demonstrated by Bouchard (1998:87) and Boucher (1999), it is reflective of the changing nature of the site's inhabitants. If we consider late century evidence from the Provincial Hygiene Inspector (ANQ, Rapport du CHPQ, 1891), the Îlot Hunt site appears to have been cleaner and better maintained than many of the city's properties, where privies were used into the late 1800s. This situation may be reflective of the socio-economic standing of the site's inhabitants and their conformity to city by-laws.

Around 1881, the site was being used by more artisans, which include a cabinetmaker, virtually no unskilled labourers, and the same continued mix of businesses (Bouchard 1998:32). It appears that the sewer and aqueduct was finally installed on rue Saint-Antoine in the late nineteenth century (Dubé 1995:47; Simoneau and Rouleau 1988:89). At this time, the privies and drains in the yard were no longer in use. It has been suggested that, as the site economy had shifted away from dependence on the declining port industries, some of the buildings on site were renovated. These include the division of the 1824 house on rue Saint-Antoine into two dwellings, and the construction and subsequent renovation of the building dating to 1875 (Bouchard 1998:93; Goyette 1999:161). The truncating of the site by rue Dalhousie as well as some of the building renovation may had reduced the need for such large privy structures. However, the renovation of the 1875, which was initially a stable, demonstrates the ongoing use and importance of such buildings to the site's occupants (Goyette 1999:162). Census data indicates that about the same number of people used the site as seen in the previous decades (Agriculture Canada 1902-1906). Artifacts found from a late nineteenth century deposit show a continued use of patent medicines and ointments. Germ theory and more scientific explanation for disease control were being promoted, at this time, but did not gain immediate acceptance, indicated by the presence of such medical treatments.

CHAPTER FOUR

ARCHAEOENTOMOLOGICAL AND ARCHAEOPARASITOLOGICAL METHODOLOGY

Archaeoentomology and archaeoparasitology are sub-fields of environmental archaeology which, when judiciously applied, have the potential to both refine and augment archaeological interpretations. The use of these approaches at Îlot Hunt will aid in reconstructing environments in and around the site, identifying background and domestic faunas, and will contribute to the interpretation of nineteenth-century medical and sanitary practices.

4.1 ARCHAEOENTOMOLOGICAL METHODOLOGY

Field procedure

Archaeoentomological samples were taken from areas 16C, 16D, 16E, and 16K. Two two-litre bags of soil were taken from every level possible. Large artifacts were removed during excavation, but otherwise the samples were taken using trowels and were not screened or processed to ensure the soil was intact in small blocks. When clear definitions could not be determined between layers, as in the excavation of the 16D200/16C26 privy structure, arbitrary ten centimetre intervals were assigned, and were sampled systematically. Levels that appeared to have a particularly high organic content such as level 16D31, the bottom level of the 16D200 privy section, were sampled intensively, as this was the sole occasion they could be accessed.

Laboratory procedure

Over forty years of analysis and many revisions have resulted in the creation of a standard technique (see Table 4.1) for palaeoentomological and archaeoentomological studies (Elias 1994:30-34, Kenward 1974:19-21; Kenward et al. 1980:8-11; Morgan and Morgan 1980a:23-24).

Not all samples taken were processed. Within the privy structure some layers were omitted as it was felt that processing all layers, many of which were designated arbitrarily, would not significantly increase our knowledge of the privy system. When several samples were taken from a single layer, a portion of these was placed in storage in case future re-examination was needed. Where possible, a minimum of two litres was processed for every layer selected. Those omitted from processing were selected according to their stratigraphic position. In arbitrary columns of layers, as seen in the 16C26 section of the privy, care was taken to ensure that every other layer was processed, resulting in sufficient coverage of the entire privy deposit. Of the 188 samples taken, 59 samples or 31 percent were placed in storage at Université Laval. The remaining 69 percent or 129 samples processed weighed 293 kg and amounting to 316 litres.

Treatment of samples took place at the Laboratoire d'archéologie historique de l' Université Laval, and the methodology did include soaking all samples in washing soda before processing. Identifications were undertaken at the Canadian National Collection of Insects, Arachnids and Nematodes housed at Agriculture Canada in Ottawa, Ontario (CNC). A representative sample was taken of any non-Coleopteran remains, such as fly puparia.

4.2 ARCHAEOPARASITOLOGICAL METHODOLOGIES

The three methodologies used in processing archaeoparasitological samples are flotation, dilution, and the palynological technique. These will be discussed briefly.

Flotation technique

In the flotation technique, samples are disaggregated and then a dense solution of zinc sulphate, zinc sodium, zinc chloride or zinc bromide, is added. Parasite ova or eggs float to the top of the solution and attach themselves to a glass cover slip placed on top of the container. Ova observed on the cover slip are counted and the number multiplied by a factor calculated from a scan of the non-floating residue. This provides a value for the concentration of ova in the deposit (Driscoll 1995:2; Jones 1982:68; Reinhard 1989:271).

Dilution technique

In the dilution method, (Table 4.2) the samples are disaggregated, filtered through large and small filters, and then small aliquots are examined microscopically.

By following specific proportions outlined by Jones (Jones 1982:68) the number of *Trichuris* eggs per gram are calculated (Driscoll 1995:2). This method determines the faecal content of the soil, as soils containing relatively high proportions of faecal remains are necessary to draw conclusions about hygienic and sanitary practices. According to Jones' criteria which are accepted by some parasite workers (Driscoll 1994, 1995; Reinhard 1992), samples with over 20,000 ova per gram (opg) are considered primarily faecal in nature, 2,000 to 19,999 opg are almost certainly faecal samples though they do contain other inclusions. Those with 500-1999 opg are likely from contexts with a substantial amount of faeces, while 200-499 opg have faeces as part of the background fauna, and finally, any samples with an opg of less than 200 has little significance in archaeoparasitological interpretations (Jones 1985:112).

Trichuris trichiura is considered the indicator species of the dilution method as it is naturally more abundant than *Ascaris lumbricoides* and is therefore more commonly encountered in archaeoparasitological analyses. The dilution method may be used as a first step in isolating samples rich in faecal remains. Then, if found to be appropriate for a more detailed archaeoparasitological analysis, they may be processed further (Driscoll 1995:2). As this method involves no chemical processes, the slides examined give a realistic picture of egg preservation. While the *Trichuris* or whipworm eggs are the primary objects of study, any other pollens, plant cells, cysts, and parasite eggs are also observed.

Table 4.1 Step-by-step methodology for archaeoentomological analysis

1.	The sample is weighed and the volume measured. The sample is then placed in a large clean bowl or wash tub able to hold several litres more than the sample volume. Sample sizes may vary, but are normally between 2-5 litres.
2.	Warm water is added to begin diluting the sample. If the sample appears to remain compacted, washing soda or sodium carbonate (50-100g) may be added and the covered sample is left to soak for up to a week.
3.	When disaggregated, the sample is repeatedly poured over a 300-micron (μ) geological sieve using cold water, until all visible soil has been removed. Any artifacts or other large items in the sample may be carefully removed at this point and set aside.
4.	The organic debris trapped in the sieve is placed in a clean bowl and an equal volume of kerosene is added and kneaded into the sample for about one minute in a fume hood.
5.	Any excess kerosene is decanted. Cold water is added to the sample filling the container or wash tub to within two centimetres of the top. The sample is left to sit for ten minutes in a fume hood.
6.	Insect remains and other organics that have risen to the surface on the water/kerosene interface are carefully decanted onto a clean 300μ sieve. This is referred to as the light fraction.
7.	The light fraction is washed with household dish detergent, to remove the kerosene, and then rinsed with water until free of detergent.
8.	Any remaining water in the light fraction is drained off. It is then rinsed with ethyl alcohol. The light fraction is then stored in a covered jar or beaker immersed in ethyl alcohol. The heavy fraction may be left to dry in the fume hood or cleaned in the same way as the light fraction and examined in alcohol if desired.
9.	The insect fragments are picked out of the light fraction under a low power microscope and stored separately in a small vial of ethyl alcohol.
10.	Representative samples of these insect remains are glued onto micro-palaeontology cards using water-soluble glue. They are normally organised at this stage into orders, families, genera, or even species.
11.	Identifications are done using comparative insect collections and taxonomic keys. When necessary, these determinations are confirmed by entomologists who specialise in the relevant insect groups.
12.	Once identified, counts of minimum number of individuals (MNI) are calculated by combining both the glued insect fragments and the pieces remaining in ethyl alcohol. Heads, pronota and elytra are normally used in this procedure.
13.	Relevant ecological data is researched in the entomological literature and is used to interpret the samples.

Table 4.2 Step-by-step explanation of the dilution technique for archaeoparasitological analysis

1.	Three grams of soil from the soil sample are added to 42 ml of water in a measured flask.
2.	This is stirred rapidly for one minute and then filtered through a 1-mm mesh to remove wood, seeds or other items of a similar size.
3.	A second filtration is then done through a 250μ mesh.
4.	While stirring rapidly to suspend the liquid, .15ml of the liquid is removed and placed in a clean tube.
5.	This entire .15 ml is examined on glass cover slips under a compound microscope and the *Trichuris* eggs are counted.
6.	This entire process is repeated three times for each bag of soil taken for parasite studies, resulting in three *Trichuris* counts per layer.
7.	Eggs for each of the three samples per layer are tabulated and multiplied by 100 to arrive at the number of ova per gram (opg).

Palynological technique

The third technique has been adapted by Reinhard (1989) and Warnock and Reinhard (1992:262-264), from the techniques used in the analysis of samples for preserved pollen remains. Small samples are screened to remove larger items and the samples are then centrifuged, treated with acid, left for a day, and then centrifuged for a second and third time with the addition of zinc bromide to separate light and heavy material in the sample. The light material forms a visible band and it is this material that is examined. Water is then used to wash away the zinc bromide and the sample left is mixed with glycerol to be examined for parasite eggs on glass microscope slides. Like the flotation technique, this approach does risk damaging the parasite ova (Hall et al. 1983:89), but has the advantage that the samples may be examined at length under the microscope unlike the dilution technique.

Archaeoparasitology methodology for Îlot Hunt analysis

Field procedure

Samples were taken in all levels opened in areas 16C, 16D, 16E and 16K for archaeoparasitological analysis. Two 100-millilitre bags of soil were taken from each level excavated, ensuring more than enough soil for the sample processing.

Laboratory procedure

The approach retained for our study combined the dilution technique and the method used by palynologists. A problematic aspect of the flotation approach is that some egg species types do not float due to their morphological structure and furthermore, application of chemicals may destroy some eggs (Driscoll 1994:48; Hall et al. 1983:88-91). In her study of latrine samples rich in parasite eggs from Colonial Boston and Maryland, Driscoll observed that the data from this technique were no more informative than the dilution method, and did not constitute a significant improvement in the quality of the data recovered (Driscoll 1994:129, 1995:7). For this reason, the dilution technique was chosen to first assess the faecal content of the soil. Further qualitative analyses could later be carried out via the palynological approach.

One of the underlying principles in archaeoparasitological studies is that samples must first be identified as being faecal in nature for archaeoparasitological interpretations to proceed (Reinhard 1992:240). In the Northeast, samples rich in raspberry seeds are considered to be indicators of deposits with a high faecal content (Fortin 1996:4; Reinhard 1992:240; 1994:63). Therefore, the first step in the examination of the Îlot Hunt material was to isolate the samples that were faecal in nature. To do so, the dried heavy fractions from the archaeoentomological analysis were examined. Samples that were visibly rich in raspberry seeds, easily observable to the naked eye, were noted. The laboratory notes from the processing of the archaeoentomological samples were also consulted as seed abundance was noted when washing down the samples. Samples that were rich in raspberry seeds were 16C55, 16C59, 16C65, 16C72, 16C75, 16D17, 16D23, 16D25, 16D27, 16D28, 16D35, 16D38, 16D41, and 16K5. They were all therefore classified as potentially containing preserved parasite eggs. These samples were all taken from within the 16D200/16C26 privy structure and its associated drain.

Samples 16C18, 16D18, and 16D26 were taken from within the 16D200/16C26 privy structure. While they did not contain many raspberry seeds, they were processed to test the raspberry seed:ova relationship at Îlot Hunt, examining whether the quantity of seed remains is in fact related to the abundance of parasite ova. The results of this testing are presented in chapter 6.

As samples for insect and parasite studies were taken from the same layers, the problem of poor disaggregation was also encountered in processing the parasite samples. Washing soda was therefore applied to the samples several days before they were processed. Two test samples were processed without its use, and the soil did not disaggregate properly, leaving clumps of soil that could not be examined in detail.

This test further ensured that washing soda did not effect parasite preservation. All samples were scanned for eggs at 200 power and all *Trichuris* eggs encountered were measured at 400 power using both ocular and objective micrometers. The presence of *Ascaris* or roundworm eggs and pollen grains were also noted during the examinations.

If the initial set of samples proved to be rich in *Trichuris trichiura* remains, according to the dilution method (Jones 1985:109), they would then be further processed to derive more qualitative information. The method chosen for this second analysis is the palynological approach (Warnock and Reinhard 1992:262-264). This method was found to be effective on studies from Israel, Philadelphia, and Newport (Warnock and Reinhard 1992:262).

Unfortunately, there was a general paucity of Trichurid remains and, based on these observations, the secondary processing using the Warnock and Reinhard method (the palynological approach) was not employed.

4.3 PITFALLS IN ARCHAEOENTOMOLOGICAL AND ARCHAEOPARASITOLOGICAL INTERPRETATIONS

Explanations using archaeoentomological and archaeoparasitological data must be undertaken with a degree of caution similar to any environmentally-based analysis used for the better understanding of an archaeological site. The following two sections provide examples of some of these difficulties experienced by other researchers.

Pitfalls in archaeoentomological interpretations

In archaeoentomological studies, Kenward has been the primary voice in discussing the interpretative difficulties in assessing insect remains (Kenward 1975a, 1975b, 1976a, 1976b, 1978a, 1978b, 1982, 1985, 1997; Kenward and Allison 1994; Kenward and Hall 1997).

One difficulty encountered is in the interpretation of the 'background noise' or background fauna of archaeoentomological assemblages. This was demonstrated (Kenward 1975a:87-88, 1976b:9) when a modern urban drain deposit was studied in an area where 95% of the ground cover was pavement or buildings. Contrary to this, the insect fauna indicated a cleared landscape with short vegetation cover around an open water filled depression. They also indicated abundant wood nearby as well as the dung of herbivores. Certain species in this assemblage were therefore over represented, likely as a result of flight, passive transport, bird droppings, and by some species living and reproducing within the drain itself (Kenward 1975a:86). If this fauna was derived from an archaeological investigation, there is a danger that the inferences made from it would be incorrect. Kenward cautions against making overly zealous interpretations in these scenarios (Kenward 1975a:87, 1976b:7) and reminds researchers to be aware that urban archaeological assemblages are usually of a very mixed origin, making interpretations difficult (Kenward 1978b:26).

Another difficulty encountered is in the habitat data available for some insect species. The value of any given insect species identified in an archaeological assemblage is only as great as our knowledge of its ecology. The available entomological literature may not cover all of a species possible habitats, and suitable microhabitats may be found in a wide range of circumstances (Kenward 1976b:8, 1978a:3, 1982:71). Such a scenario was demonstrated by the beetle species *Aglenus brunneus* thought to have been an introduction from North America to Britain. This was refuted by its presence in archaeological samples dating to the Roman occupation of England between the first and fourth centuries AD demonstrating that it is, in fact, an endemic species (Kenward 1975a:65-66; 1976a:275). Investigations using archaeoentomology have supplemented the available knowledge about this species in the entomological literature (Kenward 1975b:68).

Of further consideration for the researcher is that in archaeoentomological studies, many beetle species are consistently found together, therefore ecological information is best derived from thinking not of the implications of a single species, but from 'indicator groups' of species (Kenward 1976b:9; Kenward and Hall 1997). Studying identified beetles in ecological clusters or indicator groups provides a means to organise and interpret insect data in a manageable way. One such grouping of species commonly encountered in archaeoentomological analyses, is that of the decomposer communities, insects involved in the breaking down of dead plants and animals (Kenward 1982:71). This is a large and complex community, and divisions can be further made within this group. Some species are found with freshly foul matter while others represent an advanced state of decay (Kenward 1978a:4). In statistical studies of such groupings, Kenward warns that species that appear to be predominant in these assemblages, may just be part of the over-represented background fauna, thereby making their relationships statistically spurious or even worthless (Kenward 1982:77).

Research in Britain has also attempted to identify the 'outdoor' versus 'indoor' components of urban archaeological insect assemblages (Kenward 1978a, 1985; Kenward and Allison 1994). Isolating the 'outdoor' component may indicate whether deposits formed with more of a background fauna than would be expected for an indoor deposit. The outdoor fauna may also be studied in estimating the build-up of compost and rotting vegetation. Conversely, indoor faunas will demonstrate the importation of materials like wood into the domestic or indoor setting (Kenward 1978a:31). Outdoor insects do appear in indoor deposits. They are transported via humans, birds, other insectivores, or by their own means (Kenward 1985:101).

Two recent examples address the misuse of insect studies by archaeologists. One is seen in Cook's publication (Cook 1998) on the inhabitants of the Cross Street Back Lot site in Boston. In an attempt to reconstruct the economic standing of this household, Cook refers to rice weevils recovered in the archaeoentomological analysis and states that their presence refers to the sifting and cooking of rice, which would have been imported from Madagascar in the late 1600s. He states that the presence of rice is an indicator of good economic standing (Cook 1998:18). However, the presence of rice weevils does not unequivocally mean the consumption or even presence of rice, as rice weevils are secondary pests that may attack, in addition to rice, wheat, oats, corn, rye, buckwheat and other grain and grain products (Campbell et al. 1989:227). Cook's example demonstrates the potential for error in the imprudent use of archaeoentomological data. A second example comes from environmental studies at Harpers Ferry, West Virginia. Here, Reinhard (1994:65) suggests he has evidence for the transmission of infection by flies and water. At Harpers Ferry, faecal contamination of the water supply may have resulted in parasites in the site's human occupants, an indicator of poor sanitary practices. Poor water quality, which caused chronic diarrhoea and even death, was a common problem in nineteenth- and early twentieth-century towns. While Reinhard does not have direct evidence of this, his deduction is logical. Flies may also transmit disease and with the presence of fly puparia, Reinhard identifies a potential route of infection into the homes of those dwelling at Harpers Ferry. However, there are over 18,000 species of flies in North America (McAlpine et al. 1981:4) and without any attempt at identifying the puparia in the samples to the species level, it is a misuse of the data to imply their roles as vectors of disease.

Pitfalls in archaeoparasitological interpretations

Driscoll (1994) and Jones (1982) have made several critical observations about the practice of archaeoparasitology, primarily in terms of the lack of standardisation of the parasitological approach to archaeological soils (Driscoll 1994:2-3). Driscoll notes that there has been little consistency in methods used, no set ground rules followed for interpretations and often only one or two interpretations offered when results are considered. Furthermore, there is inconsistent use of the scientific nomenclature and little is done in the way of inter-site comparisons.

Jones also echoes this concern for inconsistency. He states that some researchers assign a specific name while other researchers are more cautious with only naming to the generic level (Jones 1982:68). A determination of *Trichuris* sp. versus *Trichuris trichiura* has significantly different interpretative value. While *Trichuris trichiura* is the whipworm known to parasitise humans, therefore significant in archaeological interpretations, an identification of *Trichuris* eggs to only the generic level may indicate parasitic infestation in dogs, mice, rats, pigs, sheep, and other mammals. Eggs of all the species in the genus *Trichuris* are morphologically similar, though they vary in size (Jones 1982:68-9). It may, therefore, be difficult to assess whether parasites are affecting human or animals, or both in groups like the Trichurids where they are morphologically so similar it can be difficult to differentiate between species (Horne 1985:307). For example, at the seventeenth century Ferryland site in Newfoundland, parasites were identified from a privy excavation. *Trichuris* sp., *Ascaris* sp., *Taenia* sp. and *Dicrocoelium dendriticum*, may all parasitise both humans and animals. The authors feel that the privy is a mixed deposit as there is a drain leading into it from what is likely a stable (Horne and Tuck 1996:514). There is no conclusive proof that the parasites were infecting humans because even *D. dendriticum*, which represents the oldest occurrence of liver fluke in New World archaeological contexts, may infest mammals such as cow, sheep and goats, which were also found at the site (Horne and Tuck 1996:514).

Health inferences from parasite studies can be tenuous at times, as the notion of health itself is a combination of nutrition, and physical well being. This may make it difficult to evaluate the real threat of many parasites (Jones 1982:68;

Reinhard 1992:238). Some parasites can have little or negligible effects on the host, while others, such as hookworm can cause disease (Reinhard 1992:239). A population that has two faecal-borne parasites such as *Ascaris lumbricoides* and *Trichuris trichiura*, is likely at risk of acquiring other faecal borne diseases (Reinhard 1992:238).

Another important consideration in archaeoparasitological interpretations is egg production, as the amount of parasite eggs recovered does not truly reflect the amount of parasitism in the population. The eggs present in faeces can vary daily depending on the number of female adult worms (Driscoll 1995:8). The female roundworm *Ascaris lumbricoides* may lay as many as 200,000 eggs per day, while other species lay as few as 50 eggs daily (Reinhard 1992:234). This may be further complicated on archaeological sites by the mixing of faecal remains from humans and animals. Driscoll suggests that due to this variation in archaeological deposits from millimetre to millimetre, we should consider the overall picture that the parasite remains presents (Driscoll 1995:8).

On a cautionary note, differences in personal habits cannot be traced through archaeoparasitology. Washing of hands and vegetables, personal sanitation habits, and clean drinking water may all potential contact with parasites. Working in contaminated areas, carrying faeces out to fertilise gardens, once a common practice, and poor personal sanitation will increase the exposure to and potential of parasitic infection (Reinhard 1994:66).

Caution must be used when utilising archaeoentomological and archaeoparasitological data to correctly assess their meaning for interpretations. However, when judiciously integrated, these approaches may challenge and improve our knowledge of past environments, health, and cultural practices.

CHAPTER FIVE

RESULTS OF ARCHAEOENTOMOLOGICAL AND ARCHAEOPARASITOLOGICAL INVESTIGATIONS

5.1 THE INSECT AND PARASITE REMAINS RECOVERED FROM ÎLOT HUNT

The insect and parasite remains discussed in this chapter were recovered from deposits on the Îlot Hunt site dating between 1850-1900, and primarily represent the fill levels of the two privies (16D25, 16D200/16C26), a drain (16D203) and a well or cistern structure (16E100) (Appendix A, Figure 18). In each section, representing the five depositional episodes identified by Bouchard (1997:33-36,38-39), the identified insect remains are described. Accompanying tables may be found in Appendix B, while a comprehensive listing of all 6755 insects analysed in this project may be seen in Appendix C. Selected images of recovered insect remains may be found in Appendix A, Figures 45-49, and they are indicated, according to their corresponding levels, throughout the text. The Coleopteran remains are listed taxonomically according to Bousquet's (1991) *Checklist of Beetles of Canada and Alaska*, and the recent revision of the Philonthina subtribe of the family Staphylinidae by Smetana (1995). European synonyms for the introduced Coleopteran species have been cited using Hansen's *Catalogue of the Beetles of Denmark* (1996).

The volume of material processed for each archaeoentomological sample (see also Table 5.1) is noted along with a short description of the inclusions found during processing. These details are a result of observations during the washing and sorting of the light fractions, along with descriptions based on the dried heavy fractions. This is followed by a discussion of the results of the archaeoparasitological analyses, which are also summarised in Table 5.2

Though not specifically the focus of this project, samples dating between 1742-1745 were processed to recover insect remains. These included 11A63, 16D3, 16C1, 16C10, 16C24, 16C27, and 16C29, all of which were found to be virtually sterile. Other samples dating to the nineteenth and twentieth century occupation of the site, including 16D1, 16D2, 16D5, 16D12, 16C9, and 16C19, were equally poor in insect remains either a result of a slight insect fauna in these contexts or poor preservation.

Archaeoentomological and archaeoparasitological remains dating to 1825-1875 (16D25)

Three samples, totalling 7.5 litres, were taken when the side of this privy was exposed during the 1995 excavations (Appendix A, Figure 21). While washing this sample, lots of raspberry seeds were noted, as well as a few small pieces of wood and mammal bones. Fragments of blue transfer-printed ware were also seen along with small pieces of iron, plaster, charcoal and brick.

Only 38 beetles were identified from this large sample (Appendix B, Table 2). Eight individuals of the Staphylinidae subfamily, Aleocharinae, were identified as well as one each of the species *Oxytelus sculptus* and *Anotylus rugosus*. As the North American taxonomy for Aleocharinae is currently under revision, beetles belonging to this group were left at the subfamily level, and no attempt was made to identify them further. Most of the beetles in this sub-family are found in animal and vegetable matter including carrion and dung, where they feed on dipterous or fly larvae and other soft-bodied arthropods (Seevers 1978:2-3). *O. sculptus* can be found in dung, decaying vegetables, mosses and compost (Hinton 1945a:54), while *A. rugosus,* one of the most abundant species of European Staphylinidae (Hammond 1976:177), appears to be a typical inhabitant of leaf litter, moss and vegetable debris in waterside settings. Most members of the staphylinid family are associated with dung or decaying organic material such as leaf litter or compost, and were likely attracted to the yard environment and the privy due to the presence of this environment.

An individual of what appears to be the black carpet beetle, also known as *Attagenus unicolor japonicus* of the family Dermestidae was identified. This introduced dermestid beetle, is currently one of the most common household pests in North America, though previous to the advent of synthetic carpets, it was even more prevalent, and likely lived in one of the dwellings using this privy. The adults feed on nectar and pollen, while the immatures or larvae may damage wool, silk, fur, skins, feathers, and other materials of animal origin, it has also been found in grain processing and storage areas, in grain dust and debris (Bousquet 1990:107; Campbell et al. 1989:246).

Ptinids, a tenebrionid or darkling beetle, a granary weevil, five lathridiid beetles and an endomychid beetle recovered from this sample, are associated with stored products. Spider beetles, represented by members of the genus *Ptinus*, are recovered in granaries and warehouses where they feed on grains and flour as well as other stored goods (Bousquet 1990:161-169). The black flour beetle, *Tribolium audax* is also found in these settings, though it does not primarily infest grains (Campbell et al. 1989:394). Also recovered was an individual of the genus *Sitophilus*, which contains the significant pest species of *S. granarius*, and *S. oryzae*, two of the most important infestors of stored grain and flour in the world (Bousquet 1990:100-101). While the ptinids, weevils and tenebrionids, may attack seeds, grains and similar products, *Mycetaea subterranea* (*hirta* Marsham, nec Schaeffer), the handsome fungus beetle of the family Endomychidae is a fungus feeder, indicating mould growth (Bousquet 1990:126). *Lathridius minutus*, another fungal feeder (Bousquet 1990:137) was also identified. As the taxonomic relationship between this and other beetles in the same genus is unclear, *L. minutus* many actually be more than one species, therefore determinations or identifications are left as *L. minutus* group. Like the endomychids, the Lathridiidae are exclusive fungal feeders, indicating their host substance is damp and mouldy (Bousquet 1990:129).

Five cucujids were identified as *Nausibius clavicornis*. This species has been found on raw sugar, dried apples, ginger and under decaying bark (Aitken 1975:128; Breese and Wise 1950:238). The reason for its presence in this privy is unclear, as it may have been associated with any of the products listed as its hosts.

Archaeoparasitological samples from 16D25 contained the highest levels of *Trichuris* ova found on the Îlot Hunt site, with the three counts yielding 300, 200 and 100 opg respectively. Some roundworm or *Ascaris* ova and pollen grains were also seen.

Archaeoentomological remains dating to c. 1850

Deposits dating to around 1850 are represented by the 16C74 deposit in the 16C26 section of the privy (Appendix A, Figure 22), and by the five layers analysed from the 16E100 structure (Appendix A, Figure 34). While 16E100 may have been previously used as a well or cistern, it was used as a place to deposit refuse around the mid-nineteenth century, and it is this deposit which is the source of the insect fauna described below.

Privy deposit 16C74 dating to c. 1850

This 9 litre archaeoentomological sample contained a rich fauna, of bugs, beetles, flies and a roach totalling 254 individual insects (Appendix B, Table 1), in its fine dry, silty soil. When processing this sample, a few raspberry seeds were noted along with eggshells, bird and mammal long bones, burned bone fragments, and fish scales. There were large wood chunks in this sample, some of which were up to 12 cm long, and during the sorting of the light fraction, there appeared to be a lot of sawdust in this sample. While no ceramics were observed, there were oxidised metal fragments, some mortar and plaster, large and small pieces of charcoal, limestone, and a piece of brick.

The members of the order Diptera and the beetles of the families Staphylinidae, Micropeplidae, Histeridae, and Hydrophilidae suggest an environment of decaying organic vegetation and dung in this deposit.

When fly (Diptera) larvae, puparium or preserved wing pieces were encountered in the light fractions they were noted and identified, though not quantified. Present were the wings of adult flies of the family Sphaeroceridae or small dung flies, larvae likely belonging to Ephydridae, puparium of the superfamily Muscoidea, and larvae of the genus *Fannia* of the Muscidae or house fly family. Three fly families make up the superfamily Muscoidea, the Scatophagidae or latrine fly maggots, Anthomyiidae, and Muscidae. The larvae of this superfamily live in a variety of habitats that include dung, plants, and animal blood (Stone et al. 1965:826, 869).

Sphaerocerids are commonly associated with all types of organic decay including dung, carrion, fungi, and dead vegetation found in damp habitats (Catts and Haskell 1990:50; Marshall and Richards 1987:994). The Ephydridae are shore flies that inhabit marshy areas, where the larvae feed on bacteria, algae, and even decaying animal tissue (Wirth et al. 1987:1028-1029). Most *Fannia* are found in mushrooms, fungi, and decaying plant and animal matter including excrement (Stone et al. 1965: 893). The presence of these four fly groups represents a damp habitat, rich in rotting organic remains, including dung.

The members of Micropeplidae family are very small beetles found in rotting vegetation. *Micropeplus tesserula*, is a Holarctic insect (found in both the Nearctic or New World and Palaearctic or Old World), which has been collected on rotting vegetables, grasses, leaves, mosses, and carrion (Campbell 1968:234). *Micropeplus browni* has been collected from old beaver lodges and from animal nests (Campbell 1968:244), and its presence here in a privy fauna suggests a previously unrecorded habitat for this native species.

The staphylinids *Oxytelus sculptus* and *Anotylus rugosus* (Appendix A, Figure 45a) were present in this sample along with a single *Carpelimus obesus,* an introduced species often found near water (Hatch 1957:91). An individual of the genus *Sepedophilus* could not be identified to the species level, though it is likely one of the introduced species from the Palaearctic region, *S. testaceus* or *S. littoreus*. The entire genus shares a similar biology. They are mycetophages or fungus feeders, and are normally collected in habitats such as rotting wood, under bark, from rotting leaves, and in decaying organic matter generally (Campbell 1976:3).

Staphylinid members of the genus *Neobisnius* are found in moist habitats, which include cultivated fields, banks of rivers, lakes and marshes. They also inhabit dung and decomposing plant material (Frank 1982:1).

Five members of the genus *Philonthus*, of the family Staphylinidae were also identified. *P. debilis* is an introduced species found in decaying organics, especially in disturbed habitats near human settlements, such as compost piles and manure. They are also found in animal nests, field and forest debris, and near water (Smetana 1995:199). *Philonthus politus* and *umbratilis* are also found in these habitats

(Smetana 1995: 76,176). Like *P. debilis*, *P. umbratilis* is found regularly near human settlements (Smetana 1995:173). *Philonthus varians* is likely an introduced species, but could not be identified beyond the species group level, though it shares the same ecological requirements as the other members of this genus. *Philonthus validus*, a species native to North America, is found in all kinds of decaying organic matter, particularly on carcasses and in dung, though they have also been collected in rotting leaves, beach litter and beach wrack (Smetana 1995:84).

The two members of the genus *Bisnius*, identified from this sample, are introduced staphylinids, which are often found near human habitations (Smetana 1995:527, 531), as is *Quedius mesomelinus* (Smetana 1971:71). These three beetles are all found in manure and decaying organic material. Similarly, members of the subfamily Xantholininae are also found in all kinds of decaying organic matter such as rotting plants, compost, human and animal dung and carrion (Smetana 1982:19).

All of the beetles identified from the family Hydrophilidae are species introduced from Europe. The abundant *Cercyon analis*, represented by 28 individuals in this sample, is found in many kinds of rotting organic matter, on windows and walls of farm buildings, and wet habitats, near water, under leaves and in wet moss (Smetana 1988:165). *C. terminatus* (Smetana 1988:158), *C. unipunctatus* (Appendix A, Figure 49a) (Smetana 1988:151), *C. pygmaeus* (Smetana 1988:157) and *Cryptopleurum minutum* (Smetana 1988:173), share these habitats.

The introduced histerid beetle, *Dendrophilus punctatus*, has been found in rotten wood, dead animals and granaries in Europe (Hinton 1945b:333), while in Canada it is often found in the cavities of deciduous trees (Bousquet and Laplante 1999a:109). In all these settings it is likely feeding on other insects as both the adults and the larvae of this family are carnivorous (Bousquet 1990:126). Larval or immature remains, represented by cephalic or head capsules of the families of either Histeridae or Hydrophilidae were recovered in the samples, providing evidence that one of these families was reproducing in this deposit.

Two *Trox scaber* from the family Scarabaeidae were identified from this level. This is a carrion feeder, which has also been found feeding on chicken feathers, owl pellets, bone meal, and frequently under dead fish and other carrion (Vaurie 1955:29). In the 16C74 deposit, *Trox scaber* may have been feeding on one of the many bones found in this deposit.

When processing this sample, many pieces of wood were observed in the heavy fraction, and five beetles represent the presence of wood in or around this deposit. The native cucujid or flat bark beetle, *Pediacus*, has been found under logs (Canadian National Collection [CNC]), while *Laemophloeus* has been collected under oak bark (CNC). The curculionid or weevil *Dryophthorus americanus* is often found on stumps under the bark of pines, and around woodpiles (Blatchley and Leng 1916:524). Two species of bark and ambrosia beetles from the family Scolytidae were identified. Nine individuals were recovered of *Polygraphus rufipennis*, (Appendix A, Figure 45d) which has white and red spruce and white pine trees as its hosts (Wood 1982:389). It attacks recently broken, cut or fallen trees (Wood 1982:386-387). *Dryocoetes autographus* is found in coniferous forests in North America and Northern Europe and Asia, and is often collected in the stumps and large roots of standing, dying trees (Wood 1982:727). Its host species in Québec are firs, white spruce, and white pine (Wood 1982:727). These beetles suggest the presence of cut trees, which are likely pines and spruces, though, oak and fir may have been present. The proximity to the port area with its shipbuilding industry, use of wooden barrel or crated by the local businesses, and stacks of wood in the yard area may have attracted these beetles.

Stored products such as grains and flour are represented by *Oryzaephilus*, a secondary pest of the family Cucujidae, which feeds on grains that have already been attacked (Campbell et al. 1989:157-160). From the family Curculionidae, *Sitophilus oryzae*, the rice weevil, shares a similar ecology. As this species is a native of tropical, subtropical and warm temperate climates, (Bousquet 1990:100), it cannot survive winters in Québec City. Members of the Ptinidae family are often found in dwellings and warehouses, (Brown 1940:121), though *Ptinus fur* is also known to feed on dried and decaying animal and vegetable matter (Follwell 1952:61). All of these stored product pests are introduced species, likely repeatedly introduced to North America during settlement, trade and immigration.

Members of the dermestid genus *Dermestes* are infestors of many stored products in Canada which include dried meats, books, pharmaceuticals, fabrics, cheese, and fruit (Campbell et al. 1989:249-254).

Mould and fungal feeders include members of the families Cryptophagidae, Endomychidae, Lathridiidae, and Monotomidae. Twelve members of the genus *Cryptophagus* and an individual of the genus *Atomaria* could not be identified to species, and were left at the generic level. The beetles belonging to *Corticaria* and *Corticarina* are also indistinguishable on the basis of the body parts used here. Like *Atomaria* and *Cryptophagus*, they are fungal feeders, which prefer moulds (Bousquet 1990:132). From the family Monotomidae, *Monotoma picipes* and, *M. testacea* have been taken in Canada in mill feeds, oat bins and decaying vegetable matter (Bousquet and Laplante 1999b:22, 35).

A single leaf beetle, from the family Chrysomelidae, *Gastrophysa polygoni*, was found in this fauna. It attacks rhubarb foliage, knotweed, bindweed and other members of the plant genus *Polygonum* which is part of the buckwheat family (Campbell et. al. 1989:98-99). It may have been feeding on a weed species or plant growing in the vicinity. The environment surrounding the privy is also represented by three species of Carabidae or ground beetle. *Bembidion versicolor* is a hygrophilous beetle usually found near small or large bodies of standing or temporary waters, on substrates that are always mixed with organic material (Lindroth 1963:377-8). *B. tetracolum* (Appendix A, Figure 46a) is usually found around refuse piles on clayey soils (Lindroth 1963:331-332), while *Trechus apicalis* is found on meagre soils, in shady settings, and amongst dead leaves (Lindroth 1963:202). These four species suggest a setting with some weeds, and organic material on a clayey or mixed substrate.

There is also an indication of some water in the area.

Two species certainly inhabited homes or buildings around the site. These are the German or domestic cockroach (Appendix A, Figure 46b) from the order Dictuoptera, and the bedbug of the order Hemiptera. The domestic cockroach, *Blatella germanica*, is the most common infesting cockroach in the world. It is an indoor pest and is usually associated with places of high humidity such as kitchens and bathrooms, but if there is adequate food and water available, will infest other rooms (Robinson 1996:141). It has also been reported from stores, warehouses, bakeries, food-processing and storage buildings, and often occurs in great numbers (Vickery and Kevan 1985:109). In Northern temperate regions, sustained infestations may occur when buildings are heated and occupied during the winter months (Robinson 1996:142).

Bedbugs are bloodsucking ectoparasites of mammals and birds. While the genus *Cimex* contains eight species, the one found here is almost certainly *Cimex lectularius*, the bedbug (Slater and Baranowski 1978:140). This species can be a great source of irritation to humans in houses or other buildings. It is transported in bedding, furniture or clothing, as it does not actually attach itself to humans when not feeding (Borror and White 1991:117).

16E100 structure dating to c. 1850

This circular structure, which previously functioned as a well or cistern, was excavated in a series of five successive layers, which became progressively richer in insect remains (Appendix B, Table 3).

A 2 litre sample was taken from layer 16E1. It was sandy and rocky and yielded a very small light fraction for sorting. A couple of raspberry seeds were seen along with several nail fragments, some woody bits, clear glass, a couple of small pieces of brick and charcoal, and chunks of schist and limestone up to 8cm in length. Unidentifiable fragments of two beetles from the families Staphylinidae and Ptinidae were the only insects recovered.

Sample 16E2 was larger, with a volume of 4.3 litres. It contained a lot of small wood fragments as well as a high volume of sand. Nails, some green glass, chunks of mortar and plaster up to 4 cm in length, limestone, schist and a few small pieces of brick were also recovered. Like the previous sample, it contained only fragments of staphylinid beetles, only, three of which could be taken to the subfamily level.

16E3 was a sample of 6.25 litres which contained brown and green glass fragments, small charcoal fragments and very small pieces of stone and wood. The insect fauna from this layer contained 14 individuals, nine of which were staphs of the subfamilies Aleocharinae and Staphylininae. Single occurrences of the hydrophilid *Cercyon,* the scarabaeid *Trox scaber,* the monotomid *Monotoma*, and Ptinidae made up the remainder of the fauna, suggesting a slightly more organic level than 16E2, along with carrion, fungal and stored product feeders.

The following layer, 16E4 was 4 litres in volume, and contained pieces of wood greater than 11 cm in length, a feather fragment, charcoal, stones, and pieces of blue transfer-printed ware. Thirty beetles were identified from his level, including 10 staphs and a histerid beetle, as well as histerid or hydrophilid larvae. *Trox scaber*, a carrion-feeding beetle was recovered, as well as three fungal feeders. The presence of two ptinid beetles as well as *Tenebrio molitor* of the family Tenebrionidae conceivably indicates the presence of stored products. Also known as the yellow mealworm, *T. molitor* has a preference for feeding on moist and decaying grain and cereal products (Bousquet 1990:190).

The fifth and final layer of this circular structure, 16E5, was sandy and contained both calcined and unburned bones, which were highly fragmented. Also recovered from this 4 litre sample, were pieces of bark, earthenware, and brown transfer-printed ware. A few square nails were also recovered along with green and clear glass, charcoal, mortar, and plaster. A large piece of wood, resembling the end of a beam measuring 23 cm by 26 cm by 50 cm, was found in this, the bottom level of the structure during the archaeological excavations (see section 3.2).

The beetle fauna from this level is similar to 16C74 in that it contains a mix of species which inhabit decaying organic remains and dung, wood, fungus, and stored products, while two ground beetle species represent the surrounding environment. Only a couple of stored product species were identified, one of which, a ptinid, *Tipnus unicolor,* is widely distributed in Europe (Brown 1940:119), and may have fed on animal and vegetable residues, decaying grain, flour, or wood (Bousquet 1990:156). The histerid *Carcinops pumilio* lives in a variety of settings including rubbish, carrion, human excrement, bark, cereals, stored and waste grain, henhouses and stables (Bouquet and Laplante 1999:118; Hinton 1945b:337).

The Carabidae or ground beetle *Pterostichus corvinus* is known to be very hygrophilous, and is found in grasses at the border of small standing waters (Lindroth 1966:499-500). It likely fell into the circular structure while scavenging the surrounding area. *Trechus apicalis* is often found on shady soils amongst dead leaves (Lindroth 1963:202). Two species of the family Curculionidae, a single *Cossonus americanus* collected beneath the bark of dead trees (Downie and Arnett 1996:1649), and 63 individuals of *Dryophthorus americanus* (Appendix A, Figure 47b) indicate the presence of wood.

Archaeoentomological and archaeoparasitological remains dating to c. 1860

16D200 layers dating to c. 1860

Nine layers from the northern part of 16D200 (Appendix A, Figures 25, 28), excavated in 1993 and 1995, (11C29, 11C31, 11C33, 16D27, 16D31, 16D33, 16D35 16D40, and 16D41), yielded 2746 insects (Appendix B, Table 4), or 41 percent of the total insect fauna identified. As the nine samples represent a depositional episode dating to around 1860 they will be discussed as one fauna.

Samples 11C29, 11C31, and 11C33 were processed and analysed (Bain 1997a) in 1993, primarily to test the suitability of this privy for further research. The same detailed procedures of recording their volumes and inclusions were not undertaken at this time, and are not available, however, the insect fauna has been integrated into this study.

The archaeoentomological samples 16D27 (16 litres), 16D31 (24 litres), 16D35 (8 litres), and 16D41 (10 litres) contained similar inclusions. While 16D27, 16D35 and 16D41 were rich in raspberry seeds, 16D31, the largest sample processed in this project, did not contain raspberries, but instead contained coffee beans. All four samples were rich in eggshells, fish scales, mammal, bird, and fish bones. They also contained bark fragments, and small and large pieces of wood up to fifteen centimetres long, some of which were the products of woodworking. Other inclusions included mortar, plaster, limestone, and lots of brick and charcoal fragments. The artifacts found in these samples included pipe fragments, blue transfer-printed ware, a slate crayon, pieces of metal and copper, clear and dark blue glass fragments, two buttons, leather and fabric pieces, as well as some glass beads.

Samples 16D33 and 16D40 were smaller samples of 2.2 litres and 4 litres respectively. Sample 16D33 yielded a very small light fraction and contained a lot of charcoal. Other inclusions were a few fish vertebrae and scales, as well as mammal bones, a few pieces of flown blue ceramics, some mortar and plaster, schist, and a brick fragment. Sample 16D40 contained a lot of sand and wood fragments, some of which were large with their bark still attached. Oxidised metal fragments, green glass, and few fragments of mortar and plaster were also noted while processing this sample.

Four of the recovered beetles, which include the staphylinid *Creophilus maxillosus* (Appendix A, Figure 47c), *Necrodes surinamensis* of the family Silphidae (Appendix A, Figure 47d), *Omosita colon*, of the Nitidulidae family and the scarabaeid *Trox scaber* (Appendix A, Figure 47a) are carrion feeders. *N. surinamensis* is a widely dispersed carrion feeder, generally found on larger carcasses such as dog, deer, bear, and moose which it uses for reproduction (Anderson and Peck 1982:32). In studying this species, Ratcliffe found that it arrives on carrion in the early stages of decay and adults feed on both maggots and on the flesh of the carrion once maggots have buried themselves (Ratcliffe 1972:390). *Omosita colon* is an introduced carrion feeder that has been occasionally found in dwellings and empty granaries (Bousquet 1990:154). *Creophilus maxillosus* is a large carrion-feeding staphylinid beetle, though the geographic origins of this beetle in this sample are unclear. *C. maxillosus* has both a native North and Central American population, and an introduced Palaearctic population, both of which currently exist in North America. The three individuals identified in this study did not have the necessary characters intact to separate them to the proper subspecies of *C. maxillosus maxillosus* (Linné), which is introduced, and *C. maxillosus villosus* (Gravenhorst), the species native to North America (Smetana 1965:33-34).

All nine levels contain many species that suggest an environment with decomposing organic remains and dung. The previously discussed Diptera, *Micropeplus* of the family Micropeplidae and the staphylinids, Aleocharinae, Xantholininae, Staphylininae, *Philonthus, Quedius, Carpelimus, Sepedophilus, Oxytelus, Anotylus* and *Bisnius* beetles, as well as the members of the family Hydrophilidae found in these levels, all share this habitat.

The native staphylinid species *Syntomium grahami* is normally found in settings with sphagnum moss and the collecting data on this species (CNC; Smetana 1988:45-46) would suggest it has a preference for cool, damp habitats. While there is no North American habitat data for *Philonthus discoideus*, of the same family, in Europe it is found around human settlements, in compost, rotting vegetation and in the dung of sheep, cows and horses (Smetana 1995:189), as is the introduced staphylinid *Gyrohypnus angustatus* (Smetana 1982:195). *Cilea silphoides* is another introduced staphylinid beetle widely distributed in the Palaearctic region, that is normally found in piles of rotting vegetable matter, such as compost heaps, piles of grass, fruit or straw, dung and old mushrooms (Campbell 1975:203). *Erichsonius pusio* has been collected from oak-beech forest litter (Frank 1975:190). Records from the CNC show it has been collected from tree litter near water, and on a mossy log in a pond, suggesting it favours a wet, highly organic habitat. *Neohypnus obscurus* is a native species and is the most commonly found species of the Staphylinid subfamily Xantholininae in some settings (Smetana 1982:227). It occurs in a wide variety of habitats such as around moss, leaf litter, floor debris in coniferous and deciduous forests, compost, mouldy wood chips, drying cow dung, and carrion (Smetana 1982:227). The substrate that attracted *Hapalaraea floralis* (=*Phyllodrepa floralis* [Paykull]) is less clear. As its name implies, it favours flowering vegetation and herbs, though its larvae inhabits rotting vegetation (Horion 1963). The Clambidae *Clambus armadillo armadillo* (Horion 1949:171), along with the histerids *Atholus bimaculatus* (=*Hister bimaculatus* Linné), and *Carcinops pumilio* also live in decomposing organic matter (Bousquet and Laplante 1999a:97, 337).

Six species of beetles indicate the presence of mammal dung in these samples. The histerid *Margarinotus merdarius* (=*Hister merdarius* Hoffmann) lives in dung (Bousquet and Laplante 1999a:75), as does *Aphodius fimetarius*, an introduced scarabaeid beetle that feeds on cow and horse dung (Ratcliffe 1991:42). Another dung beetle, *Aphodius granarius* is not exclusively associated with dung, but it is the locale from which it is most often collected (Gordon 1983:639). The staphylinids *Philonthus ventralis* (Smetana 1995:194), and *Gyrohypnus fracticornis* (Smetana 1982:192), and the hydrophilid *Cercyon haemorrhoidalis* (Smetana 1988:155-6), and are all primarily found in dung.

Wood-damaging species recovered from these layers include the curculionid *Dryophthorus americanus, Pediacus fuscus,* and *Laemophloeus* of the family Cucujidae, and *Priobium sericeum* of the family Anobiidae, a species known to damage floors and timbers in Canada (Campbell et al. 1989:12). *Hylurgops rugipennis pinifex*, the native elm bark beetle of the family Scolytidae was also identified. It attacks elm trees in Canada and the United States and is the vector or agent in spreading the fungus that causes Dutch elm disease (Campbell et al. 1989:368). The chrysomelid or leaf beetle *Crepidodera* may also attack elms, as well as apple, pear, maple, willow, poplar and birch trees (Campbell et al. 1989:82; Parry 1986:156). *Xyloterinus politus*, another scolytid species identified, has mostly deciduous trees as its hosts, which include oak, birch, beech, ash, maples and alders (Wood 1982:746). The presence of pine in this deposit is suggested by *Sphaeriestes virescens* (Downie and Arnett 1995:1111) of the family Salpingidae.

Five scolytid beetles prefer damaged or felled trees and slash. *Pityokteines sparsus* is often found on balsam firs (Bright 1976:148), while *Orthotomicus caelatus* may ravage any species of pine, larch or spruce. This is a common species that attacks stumps, limbs, and branches of dying trees (Wood 1982:664). *Trypodendron lineatum* attacks virtually all species of fir, spruce, pine, alder, birch, juniper, and crab-apple trees, and prefers felled and injured trees larger than ten centimetres in diameter (Wood 1982:753). Species of the genus *Pityophthorus* live in a wide variety of habitats, which include slash, broken branches, and unthrifty seedlings (Wood 1982:993), while *Dendroctinus simplex* also prefers large standing trees, usually those larger than six inches in diameter (Wood 1982:150).

Many beetles in this fauna are grain and flour infestors. *Rhyzopertha dominica* of the family Bostrichidae, also known as the lesser grain borer, attacks a variety of foods including grains. Like the curculionid *Sitophilus oryzae*, it is not cold hardy and is usually found in heated buildings (Bousquet 1990:71). *Cryptolestes* of the family Cucujidae, and the tenebrionid the red flour beetle or *Tribolium castaneum*, are pests of a wide variety of stored grains and seeds (Bousquet 1990:92, 191), as is *Tribolium confusum* or the confused flour beetle (Bousquet 1990:192), though *T. confusum* primarily attacks grains (Campbell et al. 1989:397).

The curculionid *Sitophilus granarius* (Appendix A, Figure 45c) and *Oryzaephilus surinamensis* of the family Cucujidae are two of the most serious pests of stored grains. *S. granarius* is a notorious pest of stored grains and flour around the world (Bousquet 1990:100), and is often found, as seen in this fauna, with *S. oryzae*, the rice weevil (Campbell et al. 1989:227). Two members of the family Anthicidae or antlike flower beetles, *Anthicus floralis* and *Anthicus formicarius*, feed on grain residues and not necessarily on the kernels themselves (Bousquet 1990:61). *A. floralis* may also be found with dried fruit (Campbell et al. 1989:16). According to Hinton (1945b:325), the histerid *Gnathoncus rotundatus*, identified in these samples, has been found in Europe in granaries, birds' nests and carrion while in Canada it has also been taken in henhouses and flourmills (Bousquet and Laplante 1999a:146).

Dampness in grains and stored products, indicated at times by the presence of *Tenebrio molitor* (Bousquet 1990:190), can lead to fungal growth. From the family Lathridiidae, fungal feeders identified from these levels include *Lathridius minutus* group, represented in this fauna by an impressive 168 individuals, *Cartodere constricta*, *Dienerella ruficollis* and *D. filiformis* (Bousquet 1990:129, 134-135), and *Corticaria* or *Corticarina.* The endomychid *Mycetaea subterranea* (Bousquet 1990:126), and the monotomid beetles, *Monotoma longicollis* (Bousquet 1990:175) and *M. quadrifoveolata* (Bousquet and Laplante 1999b:28), as well as the foreign grain beetle, *Ahasverus advena* (Bousquet 1990:90) from the family Cucujidae share similarly mouldy habitats, as do members of the family Cryptophagidae including *Henoticus serratus* (Bousquet 1990:80).

Other infesting species, such as *Nausibius clavicornis* (Appendix A, Figure 45b) from the family Cucujidae, are more general infestors with multiple hosts. The dermestid *Dermestes lardarius* (Appendix A, Figure 49d) or the larder beetle is frequently found in Canada in granaries, grain elevators, warehouses and domestic dwellings where it feeds on all kinds of animal materials and insects (Bousquet 1990:109), and is a fairly common household pest (Campbell et al. 1989:251). The drugstore beetle, *Stegobium paniceum* from the family Anobiidae is a generalist that attacks nearly all dried plant and animal products, but especially favours bread, drugs and vegetable materials rich in starch (Campbell et al. 1989:12), while *Tenebrio obscurus,* the dark mealworm of the family Tenebrionidae, feeds on materials of both animal and vegetable origin and has been found in stables (Bousquet 1990:190)

Two species, a nitidulid and a curculionid, may suggest the presence of fruits in this deposit. *Carpophilus hemipterus* or, the dried fruit beetle, attacks both fresh and dried fruits, as well as grains such as corn, wheat, oats, and rice. It is considered a serious pest of dried fruits (Campbell et. al. 1989:316). Weevils of the genus *Curculio* feed on many types of fruits including peaches, pears, apples and apricots as well as occurring in oaks in Canada (Gibson 1969:242).

A curculionid or weevil, an elmid, and three ground beetles describe the setting in or around the privy structure. Weevils of the genus *Listronotus* from the family Curculionidae are found on a variety of plants including carrots and arrowheads which are aquatic flowers found in pond edges and shallow water (O'Brien 1981:72) while the ground beetle *Trechus apicalis* favours shady areas (Lindroth 1963:202). Another carabid, *Pterostichus corvinus* is found in the grasses at the border of small standing waters (Lindroth 1966:499-500), while *Bembidion scopulinum* is a shore-living species, found on the margins of running and standing waters on all soil types, be they barren or sparsely vegetated (Lindroth 1963:343). An Elmidae from the genus *Optioservus* was also recovered. The species belonging to this genus found in Canada all favour rapid, flowing areas of streams or currents (Brown 1972:17,19).

Insects related to human presence and their activities in this deposit, include a total of 23 *Cimex* cf. *lectularius* or bedbugs (Appendix A, Figure 46d) from the family Cimicidae, two cockroaches from layer 16D33, and what appears to be two Phthiraptera or lice from layer 16D31, though the remains were highly fragmented and cannot be identified. There also appears to be evidence of three individuals from the Meloidae family, represented by *Lytta* cf. *vesicatoria* (Appendix A, Figures 48c, 48d), popularly known as 'Spanish Fly', 'mouche d'Espagne', and 'mouche de Milan' (Mulsant 1857:159). The genus *Lytta* is a known source of cantharidin, a substance found in almost all members of the Meloidae or blister beetle family (Selander 1960:9). Cantharides were used in medical treatments since Ancient Greece, for treating tuberculosis, urinary disorders, gonorrhoea, ear and toothaches, rheumatic pains, gout, skin problems, rabies, leprosy and even hair growth deficiencies (Hakbijl and De Groot 1997:132; Selander 1960:10). Small quantities may cause bladder irritations, produce abortions, and by irritating the urinary tract, may lead to increased sexual desire, perhaps its most notorious application which continues to date (Cambefort 1994:143).

Three samples, 16D27, 16D35 and 16D41, were investigated for *Trichuris* ova. Sample 16D27 yielded 0, 200, and 0 opg, while some *Ascaris* were also noted in these samples as well as *Pinus* or pine pollen grains. The 16D35 samples contained 0, 100 and 0 *Trichuris* opg, as well as pollen grains and *Ascaris* ova. *Ascaris* and *Pinus* were also seen in the 16D41 sample, the richest of the three samples processed from this episode, with opg of 100, 200 and 100 respectively.

Privy deposit 16C70 dating to c. 1860

This 7.75 litre sample, taken for insect analysis, represents the second use of the 16C26 section of the privy. Like 16C74, it was silty in nature, though some parts of the sample remained in compacted lumps, and had a distinct ammonia odour. Inclusions were feathers, mammal bones, fish ribs and vertebrae, as well as sawdust and lots of wood which made up 60-80% of the heavy fraction. Clear glass, some anthracite, a piece of a pipe stem fragment and a piece of fabric was also recovered in this sample.

This sample contained many species which were also recovered in the c. 1860 samples from 16D200 (Appendix B, Table 4). With a total of 186 insects (Appendix B, Table 5), the Diptera and the several beetle families of Micropeplidae, Staphylinidae and Hydrophilidae, imply this layer was at least partially composed of decomposing organic remains, and does have some amount of water about indicated by the carabid *Bembidion versicolor.* Conversely, also present is *Syntomus americanus*, a xerophilous ground beetle which prefers dry conditions and sparsely vegetated settings (Lindroth 1969:1056). The CNC records this species being taken in mixed litter, forest edges, and on sandbars.

Dung related species from 16C70 include the staphylinids *Philonthus ventralis* and *Gyrohypnus fracticornis*, as well as other beetles that do occasionally inhabit dung. Several mould and fungal feeders appear in this fauna, though its stored product fauna is somewhat reduced in comparison with the 16D200 samples. The presence of wood is also suggested by these samples, though again, by fewer species, which indicate mostly pines and spruces. A single *Phyllotreta striolata* (*vittata* auct. nec Fabricius), the striped flea beetle from the family Chrysomelidae, was recovered. This chrysomelid feeds on cruciferous vegetables such as radishes, turnips and cabbage (Campbell et al. 1989:123), and indicates the presence of such produce.

The Meloidae *Lytta vesicatoria* was present (Appendix A, Figure 48d), as in the previous sample, suggesting of some sort of medical treatment, while bedbugs, domestic cockroaches (Appendix A, Figure 46b), and what appears to be the remains of a louse were also identified. Other than the bedbugs, one each of Nabidae or Damsel Bugs and Anthrocoridae were recovered. Damsel bugs are fiercely predatory insects and all species of this family feed on other insects (Slater and Baranowski 1978:134). In forensic studies, Nabidae are commonly reported from carrion, where they feed on maggots (Catts and Haskell 1990:51). The members of the family Anthrocoridae, commonly known as the Minute Pirate Bugs, are also predacious, feeding mostly on small arthropods and can be found in numerous settings such as flowers, trees, conifers, in grains and in nests of birds and mammals (Slater and Baranowski 1978:143).

16D203 drain dating to c. 1860

Three layers from the 16D203 drain, 16D30, 16D38 and 16K5, were analysed for insect and parasite remains.

Sample 16D30 was a small compact .5 litre sample taken to investigate preserved insect remains. It was difficult to dissolve, but contained some raspberry and apple seeds, and lots of small, fragmented fish, bird, and mammal bones as well as a few pieces of transfer-printed ware, clear glass and small pieces of charcoal, stones, bricks and mortar.

A second sample taken from the 16D203 drain was 16D38, found to be extremely rich in raspberry seeds. Over 100 ml of seeds were recovered from this 4.5 litre sample. Accompanying samples were also processed for *Trichuris* ova. No ova were recovered, though the soil sample for parasite processing did contain around eighty raspberry seeds, as well as apple and cherry seeds. Like 16D30, this archaeoentomological sample contained the remains of many small mammal remains, portions of which were burned. Artifacts found in the sample included green transfer-printed ware, some brown and green glass and heavily oxidised nail fragments. Over 50 percent of the heavy fraction was schist and limestone.

Like 16D38, 16K5 also contained over 100 ml of seeds that included raspberry, figs, strawberries, grapes and prunes, though the majority of these were raspberry seeds. This 10 litre sample also contained large and small mammal remains, wood fragments, charcoal, nails, and red, green and clear glass pieces. A clump of what appears to be wool was also noted while sorting the light fraction. When archaeoparasitological samples were processed for trichurid remains, 200 opg were recovered from the first count, while the other two counts produced no ova. *Ascaris* ova were also noted in the first two counts.

Sample 16D30 (Appendix B, Table 6), yielded a very small insect fauna, containing two unidentifiable Coleopteran fragments, as well as a fragment of the spider beetle *Ptinus*, suggesting poor preservation or a lack of organic material in this sample.

Samples 16D38 and 16K5 (Appendix B, Table 6), if their volumes are considered, were not particularly rich in beetle remains. However, they did contain several species of staphylinids, as well as *Cercyon* and *Micropeplus* of the families Hydrophilidae and Micropeplidae, suggesting a decaying organic component to these samples, though the only Dipterous remains recovered were from Ephydridae. There is a presence of stored product remains seen in the samples, due the presence of the Histeridae and Ptinidae species, while the monotomid *Monotoma quadrifoveolata* represents some fungal development, though it does not suggest a large presence of materials attacked by fungus. Cut wood or branches are suggested by the presence of three beetles, *Pediacus,* a cucujid, *Dryophthorus americanus* a curculionid, and the scolytid *Pityophthorus.*

Archaeoentomological and archaeoparasitological remains dating to c. 1870

These fourteen levels represent the infilling of most of the 16C26 section of the privy around 1870, and yielded a total of 2268 insects (Appendix B, Table 7) or 33.6 percent of the site's insect fauna. Due to our excavation policy of opening arbitrary levels within the privies, as stated in Chapter 3, not all samples from the 16C26 privy area were treated. After processing and sorting several samples, the insect fauna appeared to be quite consistent, and would not represent a significant addition to overall interpretations if every sample was investigated. Samples 16C44, 16C47, 16C52, 16C54, 16C55, 16C56, 16C57, 16C59, 16C61, 16C65, 16C66, 16C69, 16C72, and 16C75 were treated. Five samples, 16C50, 16C51, 16C53, 16C58 and 16C67, were not processed and remain in storage at Université Laval. Sample volumes are noted in Table 13.

The samples were fairly homogenous in their inclusions, and five of them, 16C55, 16C59, 16C65, 16C72, and 16C75, included many raspberry seeds, so accompanying archaeoparasitological samples were processed for preserved parasite ova. The archaeoentomological samples were all rich in bird, fish and mammal remains, some of which displayed butchery marks. Eggshells, hairs and a horn fragment were also recovered, as well as a lot of wood and what appeared to be sawdust. Artifacts included earthenware and white ware, a piece of a ball clay pipe stem, fragments of charcoal, brick, glass, nail fragments, limestone and other small stones. Overall this group of samples contained very few artifacts compared to the 16D200 section of this privy, but contained a higher proportion of wood and bones.

This large insect fauna is extremely rich in staphylinid and hydrophilid remains, which represent 69 percent of the individuals recovered. However, these samples contain surprisingly few *Micropeplus* individuals, a genus that is well represented in other layers. Hydrophilidae or Histeridae larvae and Diptera remains were also consistently present throughout these layers, again confirming this deposit contained composting organic remains and dung. Mammal dung was represented by the histerid *Margarinotus merdarius* and the scarabaeid *Aphodius fimetarius,* which occurred throughout most layers. The presence of *Gyrohypnus fracticornis* of the family Staphylinidae, and the hydrophilid *Cercyon haemorrhoidalis* also indicate excrement in this deposit. While only represented by a single individual on this site, *Philonthus sericans* is one of the most common staphylinid species in eastern North America, and it is found to occur on various organic matter such as animal droppings, leaf litter, compost and carcasses (Smetana 1995:323).

The carrion feeding fauna was represented by a member of the Silphidae family, likely feeding on maggots (Catts and Haskell 1990:47), as well as the nitidulid *Omosita colon* (Appendix A, Figure 48a), the scarabaeid *Trox scaber*, and *Creophilus maxillosus* of the family Staphylinidae. Some of the bones were likely deposited after butchering animals, thus attracting these carrion feeding species.

The many *Carpophilus hemipterus* from the family Nitidulidae may indicate the presence of fruit and other stored products, while the curculionid *Sitona* sp. belongs to a genus that attacks clover, peas and alfalfa (Campbell et al. 1989:214-223). There is a good representation of stored grains, cereals and flour in this depositional episode, as well as other stored products, indicated by the weevils, tenebrionids, such as *Tenebrio molitor* (Appendix A, Figure 48b), anobiids and a single bostrichid as well as some of the ptinids, like *Ptinus* sp.(Appendix A, Figure 49b) and histerids. The families that represent mould and fungal feeders are well represented consistently throughout these levels. One of these, *Lathridius minutus* (Appendix A, Figure 49c) from the family Lathridiidae is one of the most abundant insects found from this deposit.

While there is no Spanish fly in these layers, there are a few cockroaches and bedbugs present. The cockroaches, *Blatella germanica* (Appendix A, Figures 46b, 46c) of the family Blatellidae may be from structures such as sheds, since both of these insects along with the dermestid *Attagenus unicolor* likely came from the homes contributing to this fauna. Also in these samples were the remains of three lice, two of which appear to be from the family Pediculidae which contains the head and body lice of humans (Kellogg and Ferris 1915:9), while the third individual was so fragmented it could not be identified to the family level. Layers 16C61 through to 16C72 contain the entire bedbug, cockroach and lice fauna which seems to be absent from other 16C26 levels.

Many insects also suggest the presence of wood, not surprisingly, as there were many wood inclusions found in the heavy fractions from this deposit. The native tenebrionid *Neatus tenebriodes* has been found under pine bark (Downie and Arnett 1996:1095), while the cucujids *Pediacus* and *Laemophloeus*, the salpingid beetle, as well as the anobiid *Priobium*, also indicate the presence of wood along with several members of the family Scolytidae. The scolytid *Dendroctinus rufipennis* is a species that will attack all species of spruce within its range, favouring weakened or drying trees (Wood 1982:194-195), as does *Dryocoetes affaber*, a beetle also known to attack pines (Wood 1982:730). *Crypturgus borealis* has balsam fir, and red and white spruce as its hosts (Wood 1982:741-742), while elms are also represented by the native elm bark beetle. An interesting occurrence in the Scolytidae fauna is of *Xyleborus affinis*, which is normally found from Massachusetts southward (Wood 1982:831), yet presently unknown for Eastern Canada. It attacks deciduous trees and is found in tropical and subtropical areas (Bright 1976:138). This occurrence may have found its way into this deposit after infesting a wood product, such as a barrel, crate or piece of furniture manufactured from wood grown in more southerly climates. Eight histerids appear to belong to the genus *Acritus*, which has been found under the bark of sugar maples, though this could also be the introduced species, which, like other histerids is found in dung and decomposing organic matter (Bousquet and Laplante 1999a:134).

A single *Hydrothassa vittata*, a leaf beetle or chrysomelid that is found on the common buttercup (Laurent LeSage, personal communication 1999), was recovered, indicating that they may have been growing in the yard area, while the carabids *Bembidion tetracolum* and *Trechus apicalis* confirm there were organic remains lying about.

The *Trichuris* counts from the first and third archaeoparasitological samples tested from layer 16C55 yielded 100 opg, as well as roundworm ova, though none of the three counts from 16C59 yielded any parasite ova. Only the third count of 16C65 contained any trichurid remains, with an average of 100 opg, though the first and second counts did contain over 20 *Ascaris* ova and all three counts contained *Pinus* pollen. In processing 16C72, the first and third counts contained *Trichuris* ova, numbering 300 and 100 opg respectively. Both *Ascaris* and *Pinus* were noted in all three counts. The first count of 16C75 contained *Trichuris* ova at a density of 100 opg, while the first two counts each contained a few *Ascaris* ova, and some *Pinus* pollen grains were seen in the second count.

Archaeoentomological and archaeoparasitological remains dating to c. 1875

Part of the southwestern section of 16D200 was filled around 1875 and represented here by layers 16D16, 16D17, 16D28, and 16D42.

The archaeoentomological sample from 16D16 was a 5 litre sample that was found to be highly organic, though a small percentage of the sample was made up of sand and stone. The sample contained a few mammal long bone fragments as well as fish scales. There were some pieces of oxidised nails, a clear glass fragments, charcoal, mortar, and plaster, as well as a few pieces of brick. Only 13 beetles were identified from this sample (Appendix B, Table 8). Seven staphylinids, a micropeplid, fly puparium, and beetle larvae suggest some organic decomposition, while two *Trox scaber* of the family Scarabaeidae may indicate some carrion feeding. The ptinid and anobiid beetles represent stored product and wood respectively. Even though the sample appeared to have a highly organic soil, the insect remains were highly fragmented.

Both 16D17 and 16D28 were rich in raspberry remains and other ecofacts. These samples contained similar items, which included fish scales and bones, mammal bones and a lot of calcined bones. Wood fragments, pipe stem fragments, black transfer-printed ware and stoneware fragments were also recovered. Both samples were rich in charcoal, oxidised nails and other metal, and green and clear glass pieces. Level 16D17 also contained a piece of felt and a white glass button fragment as well as pieces of schist and limestone. A small 2.5 litre sample was processed for level 16D42. The heavy fraction of this sample was 90 percent wood pieces, as well as a few pieces of bone, clear glass, charcoal and a small fragment of brick.

Samples 16D17 and 16D28 were large samples of 10.5 and 6 litres respectively, though they only contained 65 and 59 beetles (Appendix B, Table 8), resulting in a low density of insects per litre (Table 5.1). It is also interesting to note that sample 16D42, with a 2.5 litre volume, was the smallest sample in this depositional episode, yet contained 297 beetles (Appendix B, Table 8). The staphylinids *Carpelimus, Oxytelus, Atholus,* and beetles of the subfamily Aleocharinae made up 72 percent of this sample suggesting a moist, decaying environment. Other than the high count from this subfamily and three genera, samples 16D17, 16D28, and 16D42 indicate some organic decomposition, a weak presence of stored product remains, a potential carrion presence, some excrement, several mould and fungal feeding species, and some wood, likely pine or spruce, indicated by *Polygraphus rufipennis* of the family Scolytidae. As the 16D42 heavy fraction was made up of around 90 percent wood, the presence of these wood related species is appropriate. This depositional episode differs from the 1860 fill layers of the 16D200 section of the privy in that there is no ground beetle fauna, as well a reduced presence of stored product remains. No cockroaches or bedbugs were recovered though there is a single individual of what is most likely *Lytta vesicatoria* (Appendix A, Figure 48d) or Spanish fly in sample 16D17.

Samples 16D17 and 16D28 indicated levels rich enough in raspberry seeds for archaeoparasitological samples to be processed. Only the second count of the 16D17 parasite sample contained *Trichuris* ova with a density of 100 opg, though roundworm and *Pinus* pollen were visible in all three counts. None of the three counts from 16D28 yielded any *Trichuris* ova, though like the previous sample, both pine pollen and roundworm were noted.

Archaeoentomological and archaeoparasitological remains dating from 1880-1900.

16D200 fill dating from 1880-1900

A depression above the 16D200 section of the privy occurred as a result of compression of the organic levels that lay below. This area was filled in the last two decades of the nineteenth century resulting in samples 16D18, 16D23, 16D24, 16D26, and 16D29.

Sample 16D18 was 6.5 litres in volume and was about 20% sand. There were a few seeds noted in the heavy fraction along with fish scales, a tooth, and large mammal bone fragments containing butchery marks. The only ceramics noted were some pieces of stoneware. Oxidised copper and nail fragments along with clear and light blue glass pieces, some of which were burned, along with fragments of wood, charcoal and brick, were recovered from the heavy fraction.

The following level, 16D23, was similar in composition though it contained a lot of raspberry seeds and more wood. In addition to the inclusions listed for 16D18, this 3 litre sample also contained half a four-holed bone button and some pieces of limestone. Level 16D24 was made up of 20% sand. This small 2.5 litre sample and 16D26, with a volume of 3.5 litres, included lots of bone remains with 16D24 containing more mammal remains, while the 16D26 content was richer in fish bones. Both samples contained limestone, brick and charcoal while 16D24 also yielded a piece of white ware, mortar, and plaster. It was also noted that 16D26 contained a higher percentage of figs and grapes than normally recovered. Sample 16D29 was a small sample of 2 litres that contained a few pieces of mammal and fish bone, some white ware and stoneware fragments, wood, a piece of a rusty nail, clear glass and some schist.

Samples 16D24 and 16D29 resulted in only a few insect remains (Appendix B, Table 9). The small fauna, totalling nine individuals, from 16D24 is unusual in that it contains only *Anotylus rugosus* from the family Staphylinidae, the family that normally dominates the Îlot Hunt insect assemblages. Sample 16D29 contained four staphylinids, out of a total of 12 beetles. The insects from samples 16D24 and 16D29 also contain traces of wood, stored product remains, and fungus, as does 16D18, a large sample that yielded 35 beetles (Appendix B, Table 9).

Overall, the Dipterous fauna from these samples was very poor, with 16D26, a rich sample of 296 insects (Appendix B, Table 9), containing the only fly remains. What appears to be larval remains of the family Ephydridae were noted from this sample. These are often found in marshy areas, feeding on bacteria, algae, and even decaying animal tissue (Wirth et al. 1987:1028-1029). Damp areas are also depicted by the presence of *Helophorus tuberculatus* in sample 16D23 and *Syntomium grahami* in sample 16D18. The other staphylinids and hydrophilids depict, at least in samples 16D23 and 16D42, a deposit rich in decomposing organic remains such as vegetable material and dung. *Aphodius fimetarius* of the family Scarabaeidae, the hydrophilid *Cercyon haemorrhoidalis*, and *Gyrohypnus fracticornis* from the family Staphylinidae confirm this presence of animal wastes, while some members of the staphylinid fauna are also occasional carrion feeders. Carrion is also indicated by the presence of the scarabaeid *Trox scaber*.

There are several mould and fungal feeding families in these samples, which may suggest the presence of dampened products such as grains or cereals. The tenebrionid *Tenebrio molitor* feeds on moist and decaying grains and cereal products, while *Sitophilus,* a curculionid, *Oryzaephilus* of the family Cucujidae and the ptinid fauna are also associated with flours and grains. From the family Ptinidae, *Niptus hololeucus* breeds in stored products of many kinds, and is found occasionally, but not commonly as a pest of households and warehouses in Canada (Brown 1940:118).

Fruits may have been in this deposit, as suggested by the nitidulid species *Carpophilus hemipterus*, and *Curculio* from the family Curculionidae, though weevils of this genus are also known to occur on oaks (Gibson 1969:242). The presence of peas or beans is indicated by the presence of a bruchid or seed beetle belonging to the genus *Bruchus* (Boroweic 1987:72).

The surrounding environment appears to be moderately dry and grassy, as indicated by the carabid *Poecilus lucublandus*, while *Hypera nigrirostris*, the lesser cloverleaf weevil of the family Curculionidae, an introduction, suggests there was likely some clover growing nearby, or clover entered the site in animal feed.

Wood is suggested in these levels by the presence of the curculionid *Dryophthorus americanus*, the cucujids *Pediacus fuscus*, and *Laemophloeus bigutattus*, the salpingid beetle *Sphaeriestes virescens* and *Priobium sericeum* from the family Anobiidae. Scolytid indicate the presence of pines and spruces, represented in these levels by *Polygraphus rufipennis* and by *Ips* (Wood 1982:668).

While the 16D18 archaeoentomological sample was not particularly rich in raspberry seeds, it was decided to process an archaeoparasitological sample from this layer to examine the relationship between seed abundance and presence of *Trichuris* ova. The three counts contained pine pollen, but no trichurid ova. The 16D23 parasite sample did not produce any *Trichuris* ova, though they did contain *Ascaris* ova and *Pinus* pollen grains. A single count was also done for 16D26. It contained no parasite ova.

Layers related to the late nineteenth-century renovation of the 1875 building

Layers 16C14 and 16C41 represent late nineteenth century renovations to the building constructed around 1875, which is located on the southern side of the property. Layer 16C14 yielded a .5 litre sample, which was primarily composed of sand. It contained a few bone fragments up to 3.5 cm long, some fish, and calcined bones, a few pieces of wood, limestone and bricks. Limestone and bricks made up almost the entire contents of 16C41, a 1.5 litre sample. It was found to be sterile (Appendix B, Table 10). While 16C14 contained the remains of five beetles (Appendix B, Table 10), the pieces were so fragmented, due to poor preservation, that three were classified simply as Staphylinidae, while the other two were left as indeterminate Coleoptera.

Layers from a late 1800s depression below 1875 building

Similar to the layers filling the depression over the 16D200 section of the privy, these three levels filled a depression below the building dating to 1875 and above the contents of the 16C26 privy area and therefore date to the late nineteenth century.

Sample 16C16, a 2 litre sample, contained some mammal bones, nail fragments, wood, clear glass, bricks and lots of stones. Samples 16C18 and 16C28, 4.5 litres and 3 litres in volume respectively, contained similar inclusions. In addition, a few seeds were noted in processing the 16C18 archaeoentomological sample. Though this was not particularly rich in seed remains, an accompanying archaeoparasitological sample from 16C18 was tested for trichurids, though no eggs were visible in any of the counts.

The archaeoentomological samples, not particularly rich in insect remains (Appendix B, Table 11), contained the many staphylinid and hydrophilid species, the Histeridae or Hydrophilidae larvae, and a few Dipterous remains amongst others, found to occur in most samples, suggesting a deposit with decomposing organic detritus. Mammal dung is also represented by the presence of the hydrophilid *Cercyon haemorrhoidalis*, the histerid *Margarinotus merdarius*, and *Aphodius fimetarius* of the family Scarabaeidae. Some carrion may also be present as implied by another scarabaeid *Trox scaber* and the nitidulid *Omosita colon.*

These samples also contain a slight presence of fungus represented by members of the families Lathridiidae and Monotomidae, while dampened grains are suggested by *Tenebrio molitor* of the family Tenebrionidae. While there is some depiction of stored products and fungi, these are a smaller component of the fauna than normally seen on this site. The presence of wood is well represented by the commonly occurring scolytid *Polygraphus rufipennis*, the Cucujidae, Curculionidae, and two *Priobeum sercieum* individuals from the family Anobiidae. The ground beetle fauna only indicates organic wastes, though the introduced hydrophilid *Helophorus grandis* (*aquaticus* auct. nec Linnaeus) suggests the presence of temporary pools which may have been muddy or weedy (Smetana 1988:21). While there are no bedbugs or cockroaches in these samples, what is likely *Lytta vesicatoria* of the family Meloidae (Appendix A, Figure 48d) has been recovered from 16C18.

5.2 SUMMARY OF THE ÎLOT HUNT INSECT AND PARASITE FAUNAS

This project yielded insect and parasite faunas which aid in our understanding of the Îlot Hunt site and its occupants. Insects were identified from five different orders, resulting in the determination of 105 species from 35 families. An illustration of the most abundant Îlot Hunt insect families, each representing two percent or more of the total fauna, is shown in Figure 5.1. Similar to other North American archaeoentomological studies based on privy excavations (Bain 1997b, 1998a, 1998b), the family Staphylinidae clearly dominates this assemblage at 43 percent of the total fauna. The habitat created by privies is amenable to many species of Staphylinidae, as well as members of the Hydrophilidae family, amongst others. Some species within these families may establish themselves within the privy and reproduce there, as indicated by the presence of larvae belonging either to Hydrophilidae or Histeridae. Only seven percent of the fauna could not be attributed to the family level. This was largely due to the disarticulated nature of the preserved insects, making it difficult to assign loose elytra, heads or pronotal fragments to the family level, as members of many insect families are indistinguishable in this state.

Figure 5.1 Most abundant Coleopteran families from Hunt site archaeoentomological investigations

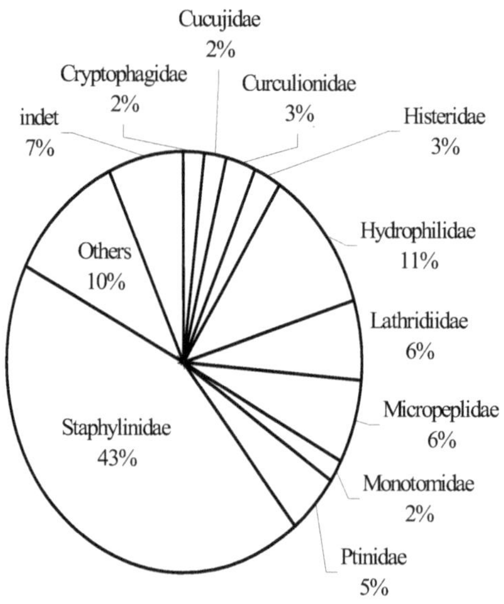

It must be pointed out that some insects occur in greater numbers than others naturally, and species abundance is not always of great interpretive value, except when it indicates probable reproduction in a layer, a phenomenon referred to as superabundance, which will be discussed in chapter six.

The bark beetle, *Polygraphus rufipennis* is generally more abundant in nature (Wood 1982:387), therefore we consistently find more of them, than other scolytid species. Conversely, Carabidae do not appear in samples as frequently as members of many other families, though their presence is of equal interpretive value.

The number of insects recovered in the layers studied for this project is not necessarily tied to sample size, clearly demonstrated in Table 5.1. At the Îlot Hunt site, insect density per litre varies widely between 146 insects/litre to only one insect/litre. This may be a result of preservation within layers, the contents of the layer themselves and the fecundity of the beetles.

A further observation on the insect fauna is the proportion of native species to those that are introduced, as illustrated by Figure 5.2. Of all the insects that could be identified to the species level for this investigation, 79 percent of all species are non-native or introduced, while 20 percent are native, and only one percent are Holarctic. Of the native fauna, half of the 825 insects belong to one species, *Micropeplus browni*, while an additional 17 percent belongs to the weevil, *Dryophthorus americanus*, thus 67 percent of the native fauna is made up of only two species.

Many species would have been introduced repeatedly during the colonisation and settlement of North America. Port settlements such as Québec City, therefore, contain faunas dominated by European species (see Lindroth 1957 for discussion), an entomological observation which has been

Figure 5.2 Introduced, native, and holarctic insect faunas from Îlot Hunt

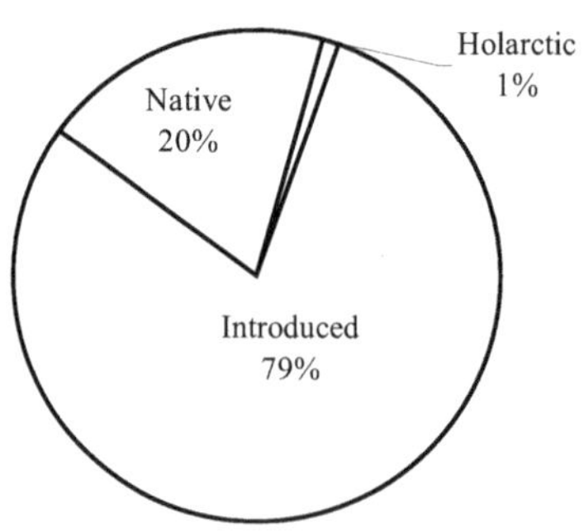

confirmed by archaeoentomological faunas from Boston and Newfoundland (Bain 1997c, 1998a, 1998b). During settlement, Europeans brought their farming systems, and with it, inadvertently transplanted many insect species, a great number of which were pests. The insects from the Îlot Hunt site which attack stored products, such as grains and cereals, are almost all introduced species. Some of them, like Sitophilus granarius, which does not fly, and is therefore tied to humans for transport, was introduced to Europe, likely from the Middle East (Buckland 1981:9), then subsequently imported to North America.

Insects living in other habitats, like the staphylinid *Quedius mesomelinus*, and the ground beetle *Bembidion tetracolum*, may have arrived in turf, which was sometimes used as ballast, in dunnage or packing materials, or in any suitable environment existing on ships making the early voyage to North America. By the first half of the seventeenth century, *Q. mesomelinus* had arrived in Newfoundland (Bain1997c:4), while *B. tetracolum* was established in Boston at least as early as 1700 (Bain 1998b:39). Often the estimated introduction dates suggested by entomologists, usually in the nineteenth century, coincide with the earliest collecting records of these species in North America, thus archaeoentomological faunas may provide the sole means of reconstructing the species' pre-nineteenth century history in North America.

Table 5.1 Archaeoentomological sample sizes and volumes

Sample	Kg	Litres	Insects/litre
11C29	1.32	-	-
11C31	.83	-	-
11C33	3.77	-	-
16C14	.71	.5	10
16C16	2.30	2	25.5
16C18	2.97	4.5	8.4
16C28	1.66	3	22
16C41	1.69	1.5	0
16C44	2.86	4	30.2
16C47	5.01	6	20.8
16C52	3.93	4	73.5
16C54	1.25	2	111
16C55	4.09	6	48.2
16C56	.47	0.5	146
16C57	.44	0.7	105
16C59	2.57	4	93.3
16C61	5.28	8	21.9
16C65	2.18	3	31
16C66	6.83	10	26.2
16C69	1.54	2	37
16C70	4.77	7.75	24
16C72	5.50	5	30.4
16C74	7.01	9	28.2
16C75	4.42	9	9.3
16D16	1.56	5	2.6
16D17	7.01	10.5	6.2

Sample	Kg	Litres	Insects/litre
16D18	5.18	6.5	5.4
16D23	2.82	3	33.3
16D24	2.35	2.5	3.6
16D25	6.5	7.5	5.1
16D26	2.30	3.5	84.6
16D27	10.57	16.0	36.9
16D28	5.36	6.0	9.8
16D29	1.89	2	6
16D30	.36	0.5	6
16D31	26.83	24	37.29
16D33	1.61	2.2	4.1
16D35	9.23	8	34.8
16D38	3.44	4.5	5.5
16D40	2.37	4.0	9
16D41	7.44	10	11.7
16D42	1.49	2.5	118.8
16E1	2.24	2	1
16E2	3.53	4.33	.93
16E3	5.46	6.25	2.2
16E4	3.51	4.00	7.5
16E5	4.23	4	29.6
16K5	8.25	10	2.2
Total	**198.93**	**238.93**	

Summary of the Îlot Hunt archaeoparasitological analyses

The results of the archaeoparasitological analysis from the Îlot Hunt site are summarised in Table 5.2, ordered according to sample and depositional event. The circa 1850 layers of the site have the highest levels of ova per gram, indicating the most faecal content of all the samples analysed at the site, while the late nineteenth-century samples examined contain no trichurid remains.

Almost all the samples studied contained some *Ascaris* ova, normally found with *Trichuris*, and pollen grains, most of which were from *Pinus* or pine pollen. As raspberry seeds are considered indicators of faecal content in archaeological soils, they were also noted during the processing of the archaeoparasitological samples. These results will be discussed further in the following chapter.

Table 5. 2 *Trichuris* counts from the Îlot Hunt archaeoparasitological analyses

Event	Sample	1st*	2nd	3rd	*Ascaris*	Pollen	*Rubus* seeds
c. 1825-1875	16D25	300	200	100	X	X	20, 75, 50
c. 1860-16D200	16D27	0	200	0	X	X	0, few, 2
	16D35	0	100	0	X	X	0, 0, 0
	16D41	100	200	100	X	X	2, 5-10, 5-10
c. 1860-16D203	16D38	0				X	80
	16K5	200	0	0	X	X	60, 60, 40
c. 1870-16C26	16C55	100	0	100	X		few
	16C59	0	0	0		X	1, 1, 0
	16C65	0	0	100	X	X	0, 0, 0
	16C72	300	0	100	X	X	0, 0, 0
	16C75	100	0	0	X	X	0, 0, 1
c. 1875-16D200	16D17	0	100	0	X	X	200+, 200+, 5
	16D28	0	0	0	X	X	2, 0, 1
c. 1880-1900- 16D200	16D18	0	0	0		X	0, 1, 0
	16D23	0	0	0	X	X	1, 2, 2
	16D26	0					3
c. 1880-1900- 16C26	16C18	0	0	0	X	X	0, 0, 0

* counts are expressed as ova per gram of soil

CHAPTER SIX

DISCUSSION

Before the recovered insect fauna from the Îlot Hunt site levels is interpreted, the concepts of the background noise and ecological grouping of the beetle fauna will be discussed. This is followed by an interpretation of the Îlot Hunt depositional events. The archaeoparasitological analysis will also be discussed in this chapter. The two data sets of environmental data are then combined with the documentary and artifactual data to examine levels of sanitation and the health of the users of the Îlot Hunt site.

6.1 Background noise and ecological groupings in the Îlot Hunt insect fauna

Understanding and properly assessing the faunas, or, the collective insect life found in these habitats (de la Torre-Bueno 1978:100) or levels, from the depositional events at the Îlot Hunt site involves distinguishing their many potential origins, and their division into ecological groupings. This interpretation is facilitated by first organising the insect taxa into clusters.

The background fauna

As indicated in chapter four, archaeoentomological faunas contain insects that may be considered 'noise'. This background noise can lead to the overvaluing of the presence of certain species, therefore, it is important that it be understood and filtered out. Insects belonging to the background noise are those that may have flown to the deposit, may be reproducing in the deposit, arrived in bird droppings, or by passive transport, via wind or water (Kenward 1975a:87-88, 1976b:9).

Some insects may be classified as overly abundant due to background noise; these are beetles that were breeding in the deposits as they accumulated. Researchers have found that compost-dwelling species make up a disproportionately large percentage of the total number of individuals in archaeoentomological faunas, indicating that they were living and reproducing there (Kenward 1978a:16). If the species makes up 10 percent or more of a deposit's fauna, then they are classified as a part of the background noise which is 'superabundant' (Kenward 1978a:16).

Ecological groupings

The creation and use of ecological groupings in discussing the identified insect faunas, is an attempt to identify clusters of insects, which represent specific habitats, and to group them accordingly for ease in archaeological interpretations. These groups of associated insects, occur as a result of their being an urban insect fauna, as they would not necessarily co-exist in more 'natural settings', yet are almost always found together in archaeoentomological analyses of town sites in Britain (Kenward and Allison 1994:55). This approach was developed by British researchers (see Chapter 4), and it has been primarily applied in England (Hall and Kenward 1990; Kenward 1976b:9; Kenward and Hall 1997), though it has also been applied in palaeoentomological studies in North America (Matthews 1983). As the Îlot Hunt layers represent mixed deposits that could be the result of a number of sources, including yard activities, the adjacent domestic and commercial dwellings with their related activities, and general urban environment of the city, it is useful to seek out these groupings for clarity in interpretations. The beetles are grouped into several categories according to ecological data noted on pinned specimens in the Canadian National Collection, as well as entomological and archaeoentomological literature (see Hall and Kenward 1990). At times, species may overlap into more than one category, though they are assigned to only one, based on their interpretive value. For example, *Listronotus* sp., is a weevil genus that is found on plants growing at pond edges, indicating a wet environment. Due to its association with water, this beetle was placed within the category of hydrophilous species, instead of the plant-associated or phytophagous beetles.

Decomposer fauna

Many beetles, particularly from the families Micropeplidae, Staphylinidae, Hydrophilidae and Histeridae, live in habitats which may include plant debris, carrion, dung of large mammals, moss, leaf litter, compost, stored product remains, burrows or nests, wood, bark and mushrooms. If we consider Table 6.1, constructed from collecting data and a survey of the entomological literature, it is apparent that most of the beetles from these four families may live in several environments and are rarely host specific. However, if we consider their preferred habitats, we can divide this data into ecological groupings as seen in Table 6.2, and therefore seek out the specific information they may be indicating. When a beetle exists in several preferred settings, as seen with *Bisnius cephalotes,* it does not aid in reconstructing the site's fauna, and is therefore considered a general decomposer. However, if a species has a marked preference for a single habitat, as seen with *Margarinotus merdarius*, they suggest the presence of a specific ecological niche. By combining the beetle preferences found in Table 6.1 with the habitat data of other beetles identified in the fauna, twelve ecological groupings within the entire insect fauna can be seen, as presented in Table 6.2.

Table 6.1 Habitat requirements of Micropeplidae, Staphylinidae, Hydrophilidae and Histeridae identified at the Îlot Hunt site.

Species	Plant debris	Wet habitats	Carrion	Mammal dung	Moss	Leaf litter	Compost	Stored products	Nests or burrows	Wood or bark	Fungi
Micropeplidae											
Micropeplus browni									√		
Micropeplus tesserula	√		X		√	√					
Staphylinidae											
Anotylus nitidulus	X	X	X	X					X		
A. rugosus	X	X			X	X			X	X	
Bisnius cephalotes	√		√	√			√		X		X
B. sordidus	√			√			√				X
Carpelimus bilineatus	X	√		X	X						
C. obesus		√									
Cilea silphoides	X			X			X				X
Creophilus maxillosus	X		√	X							
Erichsonius pusio		√			X	X					
Gyrohypnus angustatus	√	X		X		√	X				√
G. fracticornis	X		X	√		X	X		X		X
Hapalaraea floralis	X										
Neobisnius villosulus	X	√		X	X						
Neohypnus obscurus	X		X	X	X	X	X			X	
Oxytelus sculptus	X			X	X		X				
Philonthus debilis	√	X		√		X	√		X		X
P. discoideus	X			√			√				
P. politus	√		√	√	X		√		X		X
P. sericans	√		√	√		X	√		X		√
P. umbratilis	√	X				√	√		X		
P. validus	√		√	√		X			X		
P. varians group	X			X		X	X				X
P. ventralis	X	X		√			X				
Quedius mesomelinus	√			X	X	X	√		X	X	X
Sepedophilus sp.	X					X				X	√
Syntomium grahami		X			X						
Hydrophilidae											
Cercyon analis	X	X		X	X	X	X		X		
C. haemorrhoidalis	X		X	√			X		X		X
C. pygmaeus	√		X	√			√				X
C. terminatus	X			X			X				X
C. unipunctatus	X		X	X			X				
Cryptopleurum minutum	√	X	X	√			√				X
Helophorus tuberculatus					X						
H. grandis		√									
Histeridae											
Atholus bimaculatus	X			X							
Carcinops pumilio			X	X				X	X	X	
Dendrophilus punctatus			X					X	X	X	
Gnathoncus rotundatus			X	X				X	X	X	
Margarinotus merdarius				√							

***X** indicates insect has been found in this setting, while √ indicates a preferred habitat

Table 6.2 Ecological groupings of identified insects from Îlot Hunt.

Decomposer fauna: *Micropeplus browni* *M. tesserula* Aleocharinae spp. *Anotylus nitidulus* *A. rugosus* *Bisnius cephalotes* *B. sordidus* *Cilea silphoides* *Gyrohypnus angustatus* *Hapalaraea floralis* *Neohypnus obscurus* *Oxytelus sculptus* *Philonthus debilis*	 *P. discoideus* *P. politus* *P. sericans* *P. umbratilis* *P. validus* *P. varians* group *Quedius mesomelinus* Staphylininae spp. Xantholininae spp. *Cercyon analis* *C. pygmaeus* *C. terminatus* *C. unipunctatus*	 *Cryptopleurum minutum* *Helophorus tuberculatus* *Atholus bimaculatus* *Carcinops pumilio* *Dendrophilus punctatus* *Gnathoncus rotundatus* *Clambus armadillo armadillo* cf. Ephydridae Sphaeroceridae Muscoidea *Fannia sp.*
Xylophagous fauna: *Pediacus fuscus* *Laemophloeus bigutattus* *Neatus tenebriodes* *Sphaeriestes virescens* *Cossonus americanus* *Dryophthorus americanus* *Crypturgus borealis*	 *Dendroctinus rufipennis* *D. simplex* *Dryocoetes affaber* *D. autographus* *Hylurgops rugipennis pinifex* *Ips* sp. *Orthotomicus caelatus*	 *Pityokteines sparsus* *Pityophthorus sp.* *Polygraphus rufipennis* *Trypodendron lineatum* *Xyleborus affinis* *Xyloterinus politus*
Stored product fauna: *Dermestes lardarius* *Rhyzopertha dominica* *Stegobium paniceum* *Priobium sericeum* *Niptus hololeucus* *Ptinus fur* *Tipnus unicolor*	 *Carpophilus hemipterus* *Nausibius clavicornis* *Oryzaephilus surinamensis* *Tribolium audax* *T. castaneum* *T. confusum* cf. *Latheticus oryzae*	 *Tenebrio molitor* *T. obscurus* *Anthicus floralis* *A. formicarius* *Curculio* sp. *Sitophilus granarius* *S. oryzae*
Hydrophilous fauna: *Bembidion versicolor* *Pterostichus corvinus* *Carpelimus bilineatus* *C. obesus* *Erichsonius pusio* *Neobisnius villosulus* *Helophorus grandis* *H. tuberculatus* *Optioservus* sp. *Listronotus sp.*	**Coprophilous fauna:** *Gyrohypnus fracticornis* *Philonthus ventralis* *Cercyon haemorrhoidalis* *Margarinotus merdarius* *Aphodius fimetarius* *A. granarius* **Synanthropic fauna:** *Blatella germanica* cf. Pediculidae sp. *Cimex* cf. *lectularius* *Attagenus unicolor*	**Outdoor fauna:** *Bembidion scopulinum* *B. tetracolum* *Poecilus lucublandus* *Syntomus americanus* *Trechus apicalis* **Medicinal fauna:** *Lytta* cf. *vesicatoria* **Necrophilous fauna:** *Necrodes surinamensis* *Creophilus maxillosus* *Trox scaber* *Omosita colon*
Mycetophagous fauna: *Sepedophilus* sp. *Syntomium grahami* *Monotoma longicollis* *M. picipes* *M. testacea* *M. quadrifoveolata* *Ahasverus advena* *Atomaria* sp.	 *Cryptophagus* spp. *Henoticus serratus* *Mycetaea subterranea* *Cartodere constricta* *Corticaria* or *Corticarina* *Dienerella filiformis* *D. ruficollis* *Lathridius minutus* group	**Phytophagous fauna:** *Bruchus* sp. *Crepidodera* sp. *Phyllotreta striolata* *Gastrophysa polygoni* *Hydrothassa vittata* *Hypera nigrirostris* *Sitona* sp.

Coprophilous fauna

The coprophilous fauna enjoys the same environment as the decomposing beetles, though their preferred substrate is mammal dung. Many insects may inhabit dung, though in an urban archaeological setting, isolating the presence of horse, sheep and cow wastes, provides useful information about the site. Unfortunately, the same entomological information does not exist to isolate the presence of human excrement. The six coprophilous species identified in the Îlot Hunt insect fauna are the staphylinids *Gyrohypnus fracticornis* and *Philonthus ventralis,* the hydrophilid *Cercyon haemorrhoidalis*, *Margarinotus merdarius* from the Histeridae family, and the scarab beetles *Aphodius fimetarius* and *A. granarius*.

Mycetophagous fauna

Four families of beetles recovered from the Îlot Hunt site samples, Cryptophagidae, Endomychidae, Lathridiidae, and Monotomidae are all fungal feeders. When found in granaries, warehouses, or houses on stored product remains, they indicate the presence of fungal growth on the host material, and may suggest storage conditions were humid, leading to the production of moulds and funguses. However, they are also indicative of mould growth in any number of settings, as they are common or eurytopic, and not specifically associated with stored products. These beetles, as well as other species listed in this category in Table 6.2 may also be found in sheds, bark, flowers, wood, fields, mouldy hay and straw, decaying vegetable remains, and compost (Bousquet 1990:126; Campbell et al. 1989:147-150, 286-290).

Necrophilous fauna

Four beetles in the Îlot Hunt samples indicate the presence of decomposing animal remains or carrion. From the families Silphidae, Scarabaeidae, Staphylinidae and Nitidulidae respectively, these are *Necrodes surinamensis, Trox scaber, Creophilus maxillosus,* and *Omosita colon*. Occasionally these insects may be found in other settings, but they show a distinct preference for carrion.

Outdoor fauna

This categorisation includes beetles that were likely living in the yard environment at the Îlot Hunt site or out of doors generally, unlike the synanthropic fauna, which is discussed below. At Îlot Hunt this ecological grouping contains Carabidae or the ground beetles. Most members of this family are wide-ranging predators, and represent the environment or substrates immediately adjacent to the Îlot Hunt deposits. This fauna includes *Poecilus lucublandus, Bembidion tetracolum, B. scopulinum, B. versicolor, Trechus apicalis*, and *Syntomus americanus*.

Xylophagous fauna

At the Îlot Hunt site, this wood-associated group of insects includes beetles from five families. They include some of the Cucujidae, Curculionidae, Tenebrionidae, and Salpingidae, families as well as the entire fauna from the family Scolytidae which is 12 species and two genera.

Phytophagous fauna

Also associated with the urban environment is the phytophagous or plant-associated fauna, which, in the context of this discussion, includes vegetables, weeds, marsh and field plants. In the Îlot Hunt fauna, this category is represented by the chrysomelids *Gastrophysa polygoni, Hydrothassa vittata, Crepidodera,* and *Phyllotreta striolata,* the curculionids *Hypera nigrirostris,* and *Sitona* sp., and *Bruchus* sp. from the family Bruchidae

Hydrophilous fauna

The data from Table 6.1 suggests, we may consider the staphylinids *Carpelimus bilineatus, C. obesus, Erichsonius pusio, Neobisnius villosulus,* and the hydrophilids *Helophorus grandis* and *H. tuberculatus*, as species with a marked preference for wet habitats. The ground beetles *Bembidion versicolor* and *Pterostichus corvinus* and weevils of the genus *Listronotus* also inhabit such environments, as does the Elmidae *Optioservus*.

Synanthropic fauna

This small cluster of insects inhabits buildings and may be considered representative of environments where they are dependent on humans for their survival. Other species at the Îlot Hunt site, such as *Sitophilus granarius* and *Rhyzopertha dominica* could probably only survive winters in Québec if closely associated with humans in indoor settings, though in this interpretation, they are considered part of the stored product fauna. The synanthropic fauna includes the cockroach *Blatella germanica,* the dermestid carpet beetle *Attagenus unicolor* and the blood feeding Pediculidae of the lice fauna, and *Cimex lectularius,* the bedbug. It is believed that *Attagenus unicolor* or the black carpet beetle, was likely a frequent member of nineteenth century homes, more so before the use of synthetic fabrics in homes (Campbell et al. 1989:246). Robinson (1996:92) suggests that for the latter half of the nineteenth century, two of the most serious domestic pests were bedbugs and carpet beetles, thus the reason rugs were beaten. Bedbugs do not actually live on humans like lice but are nourished by human blood, but then live away from their host in suitable niches in homes (Robinson 1996:170).

Medicinal fauna

Also intrinsically linked with human behaviour at the Îlot Hunt site is the single member of the medicinal fauna which is almost certainly the blister beetle, *Lytta vesicatoria*, more commonly referred to as Spanish Fly. While our popular knowledge of this Palaearctic beetle only considers its aphrodisiac properties, documentary data indicate a widespread use for medical applications.

Stored product fauna

These species, according to the published entomological literature, as well as collecting records, are often found in warehouses and granaries, though they may exploit many types of products, including all types of foodstuffs, in a multitude of environments. They can be considered a group of 'dry' decomposers, versus the general decomposers that are found in settings with a high degree of organic decomposition, normally fouler environments (see Hall and Kenward 1990 for discussion). The stored product fauna includes the Anobiidae, Ptinidae, Bostrichidae, as well as some members of the families Scarabaeidae, Dermestidae, Tenebrionidae, Cucujidae, Curculionidae, and Nitidulidae. Disposal with household trash in floor sweepings, removal through sifting or sorting, in human or mammal excreta (Osborne 1983:460), or discarding infested materials are all means by which these beetles may have arrived in the Îlot Hunt deposits. Many insects of the stored product fauna, are occasionally, but not exclusively, found in grains and flours. Those that are strongly associated with stored grains and flour are *Oryzaephilus surinamensis, Tribolium confusum, Sitophilus granarius,* and *S. oryzae*, those these too may be found in other products at times. This sub-categorisation of these four beetles shall be referred to throughout the text as the grain infestors.

6.2 Discussion of the Îlot Hunt insect fauna

The five depositional events dated between 1850-1900 are comprised of eleven beetle faunas, which are discussed chronologically in the following section. Where appropriate, our archaeoentomological discussion is compared with observations and interpretations made by Bouchard (1997,

1998), Boucher (1999) and Fortin (1996) in their analyses.

Discussion of the c. 1850 insect fauna

16D25 privy fauna

The majority of the insect species identified from this privy sample (Appendix A, Figure 24, Appendix B, Table 2), belong to the general decomposing, stored product, and mycetophagous faunas. The decomposing beetles represented by Aleocharinae, *Oxytelus sculptus*, and *Anotylus rugosus* are likely part of the background noise and lived in the privy itself.

The *Attagenus unicolor* or the black carpet beetle, may be a species from inside one of the houses, as it was likely a common domestic dwelling species during this time period. The rest of the stored product fauna, made up of *Ptinus*, *Tribolium audax*, and *Nausibius clavicornis*, may indicate the disposal of floor sweepings or other household refuse.

The presence of *Sitophilus*, which is either the rice or granary weevil, may have arrived in this layer in human excrement (Osborne 1983:460-1), which, as a single occurrence seems the most likely route, perhaps in faecal remains in a chamber pot, one of which was recovered from this structure. Some evidence of excrement is also indicated by the presence of *Trichuris* ova, and the presence in both archaeoentomological and archaeoparasitological samples of raspberry seeds.

As sample 16D25 appears to have been taken from a level when the privy was in use for human wastes, it is not surprising to see this background fauna and the fungus feeding species, which be present due to the damp setting in this layer

The mammal bones, and seeds in the archaeoentomological samples confirm the disposal of household wastes, which included food scraps and excrement. No wood or outdoor fauna was found, suggesting perhaps this structure was closed over. The wood, plaster and brick fragments that were recovered during the processing of the archaeoentomological remains were all very small, fragmented pieces, suggesting their inclusion with other trash such as floor sweepings instead of their intentional deposition, as seen in other layers.

Level 16C74 from 16C26 privy section

This layer (Appendix A, Figure 22) was located at the bottom of the 16C26 section of the privy. As both sections of the 16D200/16C26 privy complex are contemporaneous, as indicated by the planks making up the base of these units (Appendix A, Figure 2), we may assume that rain water and other wastes flowed into the structure via the 16D203 drain. An outlet for this liquid was the 11B50/11B55 drain located at the eastern end of the 16C26 section. Hydrophilous species from this sample (Appendix B, Table 1) suggest there was water in or adjacent to this deposit, as indicated by *Bembidion versicolor*, *Carpelimus obesus, Neobisnius villosulus* and *Cercyon analis. Gastrophysa polygoni* suggests the privy was a partially open structure during the formation of this deposit, as this leaf beetle species may have been on a plant growing nearby. Conversely, in her analysis of a nearby contemporaneous level, Fortin (1996:10) suggests, by considering the small number of weed species, that this section of the privy may have been covered.

The two *Bembidion tetracolum* beetles, indicating a clayey substrate with organic refuse, and the *Trechus apicalis*, who inhabits shady settings with dead leaves, may have fallen into the privy or entered in search of food.

Polygraphus rufipennis and *Dryocoetes affaber*, amongst other Xylophagous species, indicate the presence of pine in or around this deposit. A lot of sawdust was observed in the processing of the archaeoentomological samples, likely as a result of construction or its use in the stable located south of the privy. Fragments of wood up to 12 cm long were also discarded into the privy, another potential source explaining the presence of these beetles.

Household wastes were indicated by the presence of a few seeds, burned and unburned bird, fish and mammal bones. No evidence of mammal dung is present in this fauna. The presence of grains is represented by the genus *Oryzaephilus*, and a single *Sitophilus oryzae,* while indoor or synanthropic species, represented by a cockroach, four bedbugs and a few *Attagenus unicolor* individuals. Their presence may be the result of household trash disposal, as is the stored product fauna of the ptinids, two *Trox scaber*, and a single *Nausibius clavicornis.*

The remainder of the fauna, with its many species, belongs to the two categories of decomposing and mycetophagous faunas, the majority of which are part of the background fauna, though part of this fauna may have also originated in the yard or one of the homes. As beetle larvae were found in this layer, it is suggested that we may have the establishment of a background fauna in the privy, as does the species of *Cercyon analis* which is considered superabundant in this level.

The biological content of this assemblage suggests there were some weeds and clayey soil in the adjacent area, as well, it is likely there was some wood construction taking place. Water was seemingly flowing in or near this deposit which was also the receptacle of cooking wastes and charcoal.

16E100 structure

The five layers within this well or cistern structure (Appendix A, Figures 33, 34) were the products of a rapid deposition of trash as artifacts from all layers date to the mid-nineteenth century (Bouchard 1997:57). The insect fauna (Appendix B, Table 3), supports Bouchard's interpretation. They do not have a well-developed background fauna, as seen in 16C74. Most layers investigated from the Hunt site have a strong presence of the families Staphylinidae and Hydrophilidae, which are part of the background fauna. While this structure does contain members of these families, there are no Dipterous remains, suggesting also the rapidity with which this structure was filled. There are fewer members of the general decomposing fauna suggest a drier fill than normally found on this site, which was confirmed by the presence of abundant sand and clay in most layers. There is a total lack of synanthropic, grain and dung-related faunas from all layers, though there are some stored product, mycetophagous and wood-infesting beetles present. Yard debris or general trash is further suggested by the presence of the large pieces of wood, chunks of limestone longer than eight cm, and large plaster fragments from these layers.

Overall, the layers are progressively richer in insect and cultural remains, in the lower levels of the structure, with the majority of insects in layer 16E5, which is also the only layer to contain shells and bones suggesting household wastes. If there was some water in this structure, as the beetle *Pterostichus corvinus* would indicate, heavier materials, such as refuse and artifacts may have sunk to the bottom, filtering down through the muddy levels, resulting in the most abundant insect fauna in16E5. Water may have filtered upwards via capillary action, wetting up upper layers and loosening the fill contents allowing them to sink downwards.

The 63 individuals of *Dryophthorus americanus*, 37 percent of the total fauna, from 16E5 would suggest the piece of the wooden beam disposed of in this level was made of pine and was infested, perhaps providing the impetus for its discard. *D. americanus* is an infesting species and would have been exploiting the presence of pine in this locale (Blatchley and Leng 1916:524).

These layers suggest a rapid disposal of yard trash with perhaps some household remains sinking down to the bottom layer of the structure. It was not used as a privy for the primary disposal of human and household wastes, and appeared to have been filled with general debris once it no longer served as a well or cistern.

Discussion of the c. 1860 insect fauna

16D203 drain

Three samples, 16D30, 16D38, and 16K5 were investigated from this structure (Appendix A, Figures 31, 33; Appendix B, Table 6). Sample 16D30 was a small compact, sample which contained organic remains such as seeds and bones, however, the Coleopteran remains were so fragmented they were unidentifiable except for a single *Ptinus* sp. The other two layers, 16D38 and 16K5, contained similar faunas of 22 individuals each. Both layers also contained seeds suggesting human excrement, and a diet that included apples, figs, grapes, and strawberries, as indicated by the seed remains observed during sample processing. The larger mammal, fish and bird bones were likely consumed by the privy's users, while some of the small mammal bones may have been from rodents dwelling, or being discarded in the drain. The archaeozoological evidence confirms this suggestion, as Boucher identified the bones of several rats from this feature (Boucher 1999:107).

Samples 16D38 and 16K5 contained evidence of decomposing species, a few stored product species, hydrophilid or histerid larvae, and some evidence of wood. The floral, faunal, and insect remains found in these layers suggest the discarding of excrement and some domestic wastes from cooking and perhaps chimney or stove contents indicated by the burned bones and charcoal. As they are likely fill layers to end the use of the structure, as suggested by Bouchard (1997:34), a significant background fauna was not able to establish itself within the drain. Artifact mends have been made between this structure and the 16D200 privy bottom represented by levels 16D27, 16D31, 16D35, 16D39, 16D40, and 16D41 bottom, which were rich in insect remains confirming the deposits were contemporaneous but of a different make up.

16D200 privy section

This large fauna (Appendix B, Table 4) is made up of nine layers from the lowermost levels of the northeastern area of the 16D200 privy section (Appendix A, Figures 2, 25-28). It appears that ground beetles were scavenging around this privy and may have fallen into it. They suggest clayey soils in a sparsely vegetated context, a setting not unexpected from this yard area. *Pterostichus corvinus*, *Carpelimus obesus, C. bilineatus*, *Erichsonius pusio*, *Listronotus, Optioservus,* and *Neobisnius villosulus* indicate the presence of water. *C. obesus,* may be classified as superabundant. However, none of these species are consistently present across all nine layers, but their occasional large concentrations indicate that water was entering into this layer, likely from the 16D203 drain before it was closed off, general disposal of water in the yard, or rainwater. Layer 11C29 contains a single occurrence of *Listronotus*, a beetle that is associated with plants growing in pond edges and shallow waters. If water carriers were providing water taken from shallow sources, perhaps for animals stabled on site, this may have resulted in the transport of this beetle.

There is a coprophilous fauna including *Gyrohypnus fracticornis*, *Philonthus ventralis*, *Cercyon haemorrhoidalis,* several *Margarinotus merdarius*, and both *Aphodius fimetarius* and *A. granarius*. The presence of fresh meat is suggested by *Necrodes surinamensis, Creophilus maxillosus* and *Omosita colon*, though these species are each represented by only a single individual.

Sitophilus granarius, *S. oryzae*, *Tribolium confusum*, and *Oryzaephilus surinamensis* suggest grains or flour. While these are all clear indicators of the presence of grains, a portion of the other tenebrionids, mycetophagous beetles, and the stored product insects can probably be attributed to the presence of stored grains or flour. While there was just one cockroach recovered, there were 23 bedbugs, three Spanish fly beetles, remains of two lice, and many individuals of what appears to be *Attagenus unicolor* in these layers, indicating that at least one of the adjacent households contributed to this deposit. The seed remains and other ecofacts such as eggshells and bones also suggest household trash, including human excrement, was deposited here.

The xylophagous fauna from this deposit indicates the presence of pine and elm trees as well as other coniferous and deciduous species. The ecological requirements of the wood fauna suggest there were large logs as well as cut limbs nearby. This was also evident during the processing of these samples, as wood fragments with the bark attached and pieces up to 15 cm long indicated cut marks from woodworking. This fauna may indicate the presence of a woodpile for firewood or construction, packing material, or even some infestation in the privy planks themselves.

Many species indicate general decomposition and mould and fungal growth. There is also a stored product fauna suggesting household or building wastes contributed to the formation of these layers. Evidence of superabundance by *Micropeplus browni* and large number of *Carpelimus obesus, Cercyon analis,* and *Lathridius minutus* from various levels indicate a well-developed background decomposing fauna, in a moist and mouldy environment.

These levels contained both human and animal excrement, as well as kitchen wastes, including fresh meat, which provided the necessary substrate for the carrion, dung, general decomposing, and mould faunas. The presence of plaster, wood bits, bricks and limestone also suggest general refuse disposal and some woodworking in the area, a phenomenon seen in most layers examined from this site.

Level 16C70 from 16C26 privy section

This layer represents a second use of the 16C26 section (Appendix A, Figures 2, 21) of the privy. Sandy soil existed in this deposit, as indicated by the presence of *Syntomus americanus* (Appendix B, Table 5), a beetle that favours sandy dry habitats and confirmed by the sand noted during sample processing. This beetle may have inhabited the yard area or entered the deposit with some of the many inclusions noted from this layer. Field notes from the excavation, indicate this was a clayey deposit, (Bouchard 1997:15), also indicated by *Bembidion tetracolum*. Water is indicated by the hydrophilous species, and may have flowed out of the drain east of this deposit.

These levels contained a lot of wood, which made up 60-80 percent of the heavy fraction during sample processing, while sawdust was noted during the sorting of the light fraction. The presence of wood is also indicated by several xylophagous species including *Orthotomicus caelatus*, a scolytid which attacks stumps and dying trees. If the pit was uncovered at times, perhaps the sawdust was thrown into this open pit to dry up some of the putrefying material, which still smelled very strongly of ammonia after 130 years in situ.

Two species suggest the presence of mammal dung, while grains are represented by *Sitophilus oryzae*. A phytophagous or plant-feeding species, indicated by the presence of the striped flea beetle, *Phyllotreta striolata*, suggests consumption or presence of vegetables such as radishes, turnips and cabbages, while fish and mammal bones were also recovered.

Level 16C70, from the 16C26 section of the privy, differs from contemporaneous deposits in the 16D200 section, in that there is a smaller grain infesting component, almost no seed remains, substantially more wood, and a generally reduced stored product fauna. This may indicate that there is stable and yard refuse in 16C70, yet a smaller domestic trash component, due to the smaller stored product fauna. Domestic refuse is suggested by mammal bones, the presence of the flea beetle, and by the house fauna which includes cockroaches, lice, bedbugs and the remains of what is almost certainly Spanish Fly.

Discussion of the c. 1870 insect fauna from the 16C26 privy section

The fourteen layers from 16C26 (Appendix A, Figure 2) dating to around 1870 were very similar in composition (Appendix B, Table 7), and contained a fairly uniform insect assemblage. The grain infesting species recovered in these layers was, in part, from human excrement, and from households or wastes from the grain merchant on site. While only a few species are exclusive grain feeders, there were many species present that are occasional grain infestors in the stored product fauna. Their presence in these deposits was likely in part from household wastes as well as from stable wastes. Many of the suitable niches for such species in homes, such as storage or food preparation areas, would have also been mimicked in stables. Lice, roaches, bedbugs and *Attagenus unicolor* represented the synanthropic fauna.

Bones from fish, birds, and mammals dominated the samples. While some of these were likely household trash, a proportion may have also been from the tavern on rue Saint-Antoine, an establishment which was in business under several owners throughout the entire fifty year period treated by this project. Boucher identified, via her archaeozoological analysis, many chickens, sheep, cows and pigs from this depositional episode (Boucher 1999:105-107), which would concur with this suggestion. Some of the animals were likely butchered in the yard, as suggested by Boucher (1999:102), and by the necrophilous fauna. Mammal dung was suggested by five species, again, indicating the proximity of animals being raised for consumption, or for use on site. Five of these 14 layers were rich in raspberry seeds, indicating some presence of human excrement.

The ground beetles *Trechus apicalis* and *Bembidion tetracolum* indicate a meagre clayey soil. As there were only a few ground beetles recovered from a large volume of soil, it would appear that this structure was closed over. Two beetle species represent yard plants, buttercup and clover or alfalfa, represented by *Hydrothassa vittata* and *Sitona* respectively. They may have been growing in the area or entered in animal bedding or feed remnants from the stable. The sawdust seen in many samples was perhaps discarded into the privy with stable or commercial wastes. There was a strong wood-related fauna suggesting the presence of pine, elm, and spruce in or around these levels, and many wood pieces were recovered during sample processing.

There is evidence of water in these deposits represented by abundant members of the hydrophilous insects, as well as many general decomposing and mycetophagous species. *Oxytelus sculptus* represents a superabundant species from these levels, which appear to have been mouldy and very moist to wet at times. They also contained refuse from households as well as wastes which were likely derived from the stable, work in the yard area, and the tavern on rue Saint-Antoine, represented by the presence of bones, wood, and sawdust, and the good stored product fauna.

Discussion of the c. 1875 insect fauna from the 16D200 privy section

These layers (Appendix A, Figure 30) yielded a beetle fauna (Appendix B, Table 8) that was very different from those recovered in the 16D200 layers dating to circa 1860. Bouchard (1998:70) has interpreted these events as the results of a rapid fill episode; Bouchard's hypothesis seems justified when we consider the large volume of artifacts as well as glass, metal, bone and wood recovered, indicating a lot of trash within the sample. However, the beetles indicate a well-established decomposing fauna that would imply the deposits were in situ and not immediately covered over. As there was evidence of immature species in this event, and superabundance in *Oxytelus sculptus*, *Carpelimus* sp. and the subfamily Aleocharinae, a background fauna was able to establish itself in these layers.

Quite unlike the earlier deposits, there was no outdoor fauna, no phytophagous or plant-feeding fauna, no carrion fauna, and no synanthropic fauna.

Mammal dung was indicated by a few species, and water suggested by the presence of *Carpelimus obesus*, *C. bilineatus*, and *Neobisnius villosulus*. There was a very small stored product fauna, and a smaller mycetophagous fauna than seen in the circa 1860 layers of the privy. The stored product fauna lacked some of the species normally recovered such as the ptinids, *Trox scaber* and *Dermestes lardarius* which are recovered consistently in privy faunas and were likely living in adjacent dwellings, though *Priobium sericeum*, a beetle that attacks woodwork and timbers within buildings, was present. Wood was indicated by the Xylophagous fauna, and confirmed by the wood fragments and sawdust recovered during the processing of this sample.

The sources of this fauna recovered from the 16D200 area, were large mammal faeces, household wastes, and some general trash As this section of the privy was most likely covered over, it may explain the absence of species that inhabited the yard, represented normally by ground beetles and phytophagous species. The household wastes appear to have been barely faecal in nature, indicated by the slight grain-infesting fauna. Evidence of dietary remains are indicated by the many bones recovered, some of which had been burned, indicating the disposal of stove or fireplace contents. The very small stored product fauna may be from any of the commercial dwellings, homes, or the stable, all of which would contain suitable environments. A portion of the general decomposing and the mycetophagous fauna may also hail from these environments. Other than the remains of meals, the slight faecal remains and the Spanish Fly beetle, there is only scant evidence of the interior of the Îlot Hunt homes.

Discussion of the c. 1880-1900 insect fauna

16D200 privy section

This is an interesting assemblage (Appendix B, Table 9) as it is not strictly a privy deposit. Instead, Bouchard believes it to be the infilling of depression created by compaction of lower privy layers (Bouchard 1998:73). We therefore may consider this as incidental, rather than intentional, privy use.

There was only slight evidence of wetness, represented by only two individuals of *Carpelimus*. The ground beetles also represent a dryish setting, confirmed by the abundance of sand recovered in processing these samples. A lot of wood fragments were also recovered during the processing of this sample, thus explaining the presence of the Xylophagous fauna.

While there was no carrion fauna from these layers, there was evidence of consumed household wastes, represented by the bones recovered, the pea weevil *Bruchus*, and by the presence of two members the grain infesting group of beetles, *Oryzaephilus* and *Sitophilus granarius*. *Tenebrio molitor* suggests there may have been some mouldy grains, perhaps attracting at least in part, the mycetophagous fauna. The coprophilous fauna confirms the presence of animals in the yard, likely horses or cows, while the clover leaf beetle, *Hypera nigrirostris*, may have animal feed or was present in stable litter.

General decomposers, stored product, and mycetophagous beetles were recovered, while *Micropeplus browni* was the only superabundant species. The presence of beetle larvae suggests their reproduction in this deposit as a component of the background noise. The bedbugs, and the *Attagenus* beetles imply that at least one household was a source for this fauna. Many of the stored product species were likely found in homes and were discarded in floor sweepings.

The sources contributing to these layers are general yard wastes, such as wood, domestic and perhaps stable contents. The deposits have a discernible background fauna, though they appear to have been drier than many of the lower deposit of the 16D200/16C26 privy, locales where water may have drained into or pooled regularly.

Layers related to the renovation of the 1875 building

Two layers, 16C14 and 16C41, were investigated from layers interpreted to be the evidence of renovations to the 1875 building, represented in the 1993 and 1995 excavations by the discovery of stone columns serving as the building's foundation (Appendix A, Figures 18, 20). Unfortunately, 16C41 (Appendix B, Table 10) was found to be sterile, while layer 16C14 contained only the remains of five beetles, three of which belonged to the family Staphylinidae, while the other two are unidentified Coleopteran fragments. Sand was the primary component in the 16C14 sample, and limestone and bricks made up most of 16C41, perhaps resulting in the poor preservation of the entomological remains. These results do not augment our interpretation of this time period at the Îlot Hunt site, though by their limestone and sand fills, they do appear to indicate renovations to the 1875 building.

Layers filling a depression below 1875 building

Similar to the layers discussed above in the 16D200 section of the privy dating to the late nineteenth century, these layers refer to the infilling of a depression below the building dating to 1875 (Appendix A, Figures 18, 20). These fill layers contained a few seed remains, some wood, bones, nails and stones. A single occurrence of what appears to be *Lytta vesicatoria* was recovered from this layer (Appendix B, Table 11). Its presence here appears rather accidental, as there is no other synanthropic fauna present from these layers.

Bembidion tetracolum indicates some clayey soil with organic refuse, though there are no other species that document the yard or any plants in the area. There is no synanthropic fauna, nor were there any exclusive and very few occasional grain-feeding insects present. There was some water in or around this deposit at some time during its formation represented by the hydrophilous fauna. Water running in and pooling in this deposit may have attracted *Helophorus grandis*, which favours temporary pools and muddy margins of ponds (Smetana 1988:21).

The coprophilous fauna represents the presence of large mammals in the yard. Carrion is also suggested by the necrophilous beetles. Only members of the subfamily Aleocharinae were superabundant in this layer, and the general decomposing fauna is present as usual, though there were no larval remains, and few dipterous remains recovered, suggesting a smaller decomposing fauna than normally seen from other Îlot Hunt levels.

A few species indicate the presence of wood, while the stored product species suggest some infestation of wood, stored products, and other suitable niches, which may have existed in the commercial dwellings or homes around this site.

This fauna appears to be derived from domestic trash, yard soils, and a stable fauna. The few grain related species, and lack of house or synanthropic fauna seem to suggest that domestic trash is a smaller component of these levels than recovered from the 16D200 privy layers. The presence of some stored product beetles, and the coprophilous fauna do indicate the presence of large mammals and their environment, while some wood remains, likely from the yard also contributed to this fill.

What should we consider an infestation?

Many of the species identified in the Îlot Hunt insect analysis can be classified as infestors. These include the synanthropic fauna, and the stored product fauna. When these insects, which include such unsavoury characters as bedbugs (Appendix A, Figure 46d), lice, and cockroaches (Appendix A, Figure 46b) are recovered in archaeoentomological assemblages, it is tempting to proselytise about the unclean homes and poor personal hygiene of the historic period. In comparison with our modern day standards, these homes would indeed have appeared unclean, and perhaps their inhabitants were somewhat dirty, but caution must be taken in pronouncing a home or person 'infested'.

People carrying head lice often have less than 10 on their heads, or 10-20 for body lice. Over 100 lice is rare (Busvine 1976:8). Bedbugs are about six mm long, while head and body lice are three and four mm respectively (Busvine 1976:8). To arrive in the Îlot Hunt deposits, they may have been attached to discarded clothing, or most likely, they were manually removed from hair, garments, or homes, and subsequently discarded with domestic trash. Considering the number of dwellings that may have contributed to these deposits and the number of people using these refuse disposal areas, the level of 'infestation' of these ectoparasites and cockroaches must be considered very low.

The insect faunas from the site do contain species that infest flours and grains. According to Robinson (1996:230), nine percent of all crops in the first world and over 20 percent in developing nations are currently lost due to insect infestation. We could perhaps consider loss during the historic period on par with developing nations of today, as both scenarios may lack the technology in the form of machinery and the chemical treatments so readily embraced by modern Western agribusiness. The presence of some grain-infesting or home and carpet-infesting species recovered on the Îlot Hunt site, whether they arrived in deposits in trash such as floor sweepings, stable contents, or excrement is not surprising. A bakery was located on the premises around 1850 (Leclerc 1998:27), and according to the city directories, flour merchants occupied part of the commercial buildings for every year between 1858-1871 (Cherrier 1861, 1862, 1863, 1864, 1865, 1866, 1867, 1868, 1869, 1870; Cherrier and Hamelin 1858). The arrival of new shipments would have readily replenished the stock of pest species, in both the warehouses and bakery, as many of these species are able to seek out habitable environments in adjacent dwellings. The flour and grain stores would have experienced some degree of contamination, by both primary and secondary species, and given the proximity of these businesses to the deposits studied, it is intriguing that such low numbers of stored product insects were recovered. None of the levels examined in the course of this study contained high enough levels to suggest a significant or notable infestation, with the stored product insects making up only three percent of the total fauna. In other archaeoentomological assemblages where a substantial infestation has been proposed (Bain 1998b:44 ; Kenward and Williams 1979), the stored product species made up a significant percentage of the insect fauna.

Discussion of *Lytta* cf. *vesicatoria*

There are four individuals of what is most likely *Lytta vesicatoria* (Appendix A, Figure 48d) or Spanish Fly from c. 1860, and single occurrences from c. 1870, c. 1875 and c. 1880-1900. Three of these deposits, with the exception of the 1880-1900 occurrence, also contain glass vials and medicinal bottles, which may have contained these beetles. As they were mostly small fragments recovered, they may have been discarded with the bottles. Their presence in this study is significant as it indicates the ongoing use of traditional medical treatments throughout the nineteenth century.

As it is impossible to identify the recovered fragments to species level based on the body parts recovered from the Îlot Hunt samples, which include a pronotum (Appendix A, Figure 48c), some leg and elytral fragments, some explanation of the presumed determination of *Lytta vesicatoria* should be made. Even if it is another species within the genus *Lytta*, its interpretative value remains the same for this project, as other members of this genus have the similar properties necessary for the 'medical treatments' of the nineteenth century. Other native *Lytta* species do occur in Western Canada, but it is unlikely they were in use at this time. *L. vesicatoria* is strictly Palaearctic in distribution, and is the only species of *Lytta* in Europe, thus the remains found would have been imported from Europe. They are commonly found in Southern France and Italy (Mulsant 1857:158), and could have easily been imported with other goods to a port like Québec.

As mentioned in chapter five, this was used in medical treatments ranging from tuberculosis, to rabies, to skin problems (Hakbijl and De Groot 1997:132; Selander 1960:10). Archaeologically, *Lytta vesicatoria* has also been recovered from the shipwreck 'Amsterdam, excavated by Dutch archaeologists (Hakbijl 1986, 1987). Found with the *Lytta*, was a fragment of *Cetonia aurata* or the Rose Chafer (Scarabaeidae). It has been suggested that it was used to dilute the cantharidin or may have had its own medicinal value. Another alternative explanation was the substituting of *Cetonia* for *Lytta*, as it may have been cheaper and more accessible, thus duping the purchaser.

In Duffin's (1993) biographic work on Dr. James Langstaff, a nineteenth century physician practising in Richmond Hill, Ontario, the use of Spanish Fly was noted in his diaries. Langstaff was licensed in 1849, and practised medicine for

40 years, recording his medical treatments throughout the entire period (Duffin 1993:26). Amongst his most commonly mentioned preparations during the 1850's, were plasters and poultices made of mustards and cantharides or Spanish flies, though his use of these declined later in his practice (Duffin 1993:81). These treatments would cause blistering, with the hopes of drawing the problem out of the organs and into skin (Duffin 1993:81).

Nineteenth-century Canadian recipes using Spanish flies include treatments for 'female problems', a nerve and bone liniment (Moore 1874:50), and treatment for bone felons or abscesses (Anonymous 1882:227, 1895:232). To treat bone felons, the guide suggests applying a 'fly blister' the size of a thumbnail for six hours, to draw the felon out, at which time it will be visible just below the blister (Anonymous 1895:232). The treatment for pre-menstrual problems is called the Emmenagogue Tincture or Pill, which calls for a tincture of Spanish flies (Moore 1874:50). Finally, the nerve and bone liniment calls for cayenne, amber, turpentine and a considerable volume, six ounces, of Spanish flies (Moore 1874:50).

The three recipes, and more importantly, the records of Langstaff, indicate the prevalence of Spanish Fly in nineteenth century medical applications. Bernier states that the new medical developments in Québec had little influence on medical practice before 1880 (Bernier 1989:130), and doctors were still using purges, diets, opium and morphine, old techniques which were easy to prescribe (Bernier 1989:117). The residents of Îlot Hunt were likely adhering to oft-used methods for treating their ailments.

Comparison of the Îlot Hunt fauna with the Abiel Smith School and Faubourg Québec sites

While the Îlot Hunt study is a substantially larger investigation than other nineteenth century privy studies, it is useful to compare it to other faunas to examine whether the proportions of insect families, and the percentage of introduced species are similar.

The Abiel Smith School site in Boston, Massachusetts (Bain 1998a) was the first school for African American children in the United States and was located in a densely populated sector of Boston. Samples, in conjunction with the Afro American Museum of History and the United States National Park Service, were taken from an eight-seater brick privy structure in the school yard, which dated from the mid to late nineteenth century, according to the artifactual data. Material culture used in dating the structure were recovered lamp chimney glass fragments and embossed medicinal bottles (Mead 1997:5). The privy was a large rectangular structure, 4 m long and 3 m wide, which had been divided internally, though these divisions did not extend to the bottom of the cesspit underneath, allowing for the circulation of wastes (Mead 1997:3-4). The structure was also connected to a catch basin via two drains, the slope of which indicates the catch basin drained into the privy (Mead 1997:4).

The Faubourg Québec site (BjFj-56) is a nineteenth-century military site located on the island of Montréal, near the base of rue Berri close to the St. Lawrence River. The material studied from this site was taken from structures used by civilians and included deposits from two privies and a drain dating from the early to mid-nineteenth century according to the recovered artifacts (Christian Belanger, personal communication, 1996).

The Îlot Hunt fauna (Figure 5.1, Appendix C) does have similarities to the Abiel Smith School site (Figure 6.1, Appendix B, Table 12). Staphylinidae also dominates this assemblage, and the proportions of Lathridiidae, Monotomidae, Ptinidae, and the indeterminate fauna are almost identical. While archaeoentomological faunas reflect their surroundings and deposits themselves, it is worth noting these basic similarities between the two fauna which do not appear to be unique to the Îlot Hunt site.

Figure 6.1 Most abundant Coleopteran families from the Abiel Smith School site

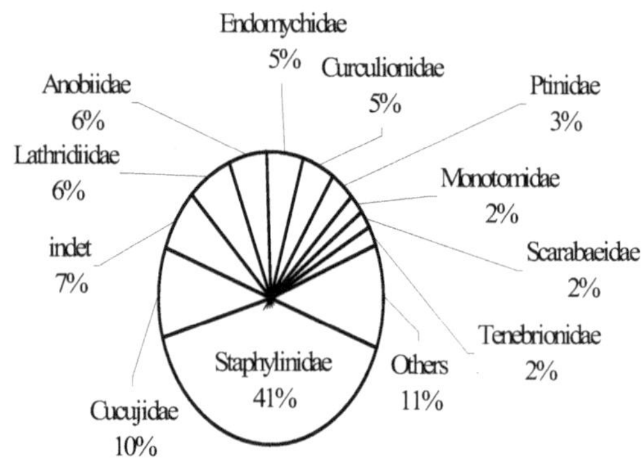

Figure 6.2 Most abundant Coleopteran families from the Faubourg Quebec site

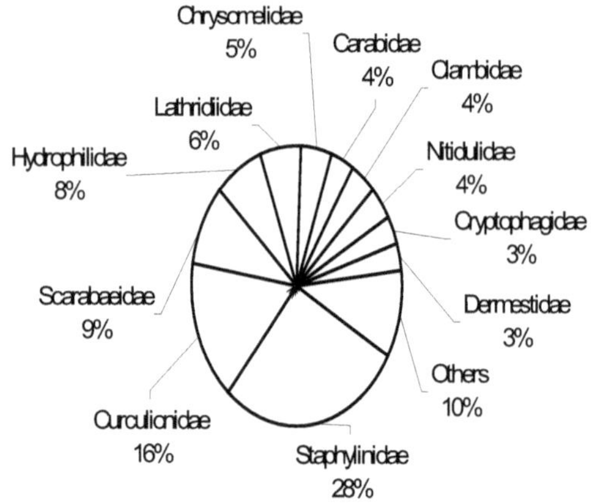

The Faubourg Québec site (Figure 6.2, Appendix B, Table 13) was also dominated by Staphylinidae, but contained greater percentages of Chrysomelidae, Carabidae, and Curculionidae which was probably an accurate reflection of the site's environment which would have been quite different from Îlot Hunt and the Abiel Smith sites. The percentage of introduced species at the Faubourg Quebec site was only 44 percent, while 60 percent of the urban fauna from Boston was introduced. Again, we can presume that due to a similar dense urban landscape, there will be more parallels between the Hunt and the Abiel Smith assemblages. By examining the tables from the three sites, the species making up the introduced component are strikingly similar.

While these are very basic and preliminary comments on comparing the Îlot Hunt material with two other Eastern North American studies, we may use this knowledge in our future studies of urban archaeoentomological assemblages.

6.3 Interpreting the parasite remains

The remains of the archaeoparasitological investigations are listed in Chapter Five (Table 5.2) by corresponding event. Immediately, it appears that the soils, from which the samples were taken, were not rich in faecal remains. Due to this, the further processing of the soil samples, using the palynological technique (see Chapter 4) was not undertaken.

As presented in Chapter 5, all *Trichuris* counts from the Îlot Hunt samples expressed as ova per gram or opg, are between zero to 300 (Table 5.2). If these are considered in conjunction with the interpretive guidelines outlined by Jones (1985), then the Îlot Hunt samples vary between 'little significance for archaeological interpretation' for samples containing less than 200 opg and, 'faeces are part of the background fauna' for those with 200-499 (Jones 1985:112).

The highest levels of *Trichuris* eggs seen were in the 16D25 privy (Table 5.2), indicating that at least part of the privy contents were, in fact, made up of faecal remains. All three counts from 16D25 contained *Trichuris*, only seen in one other sample, 16D41. The 16D25 samples also contained ova of *Ascaris* or roundworm and pollen grains. A chamber pot discarded into the privy may have contained faecal remains, though there is not a consistent association of chamber pots and faecal remains throughout the Îlot Hunt depositional events.

The circa 1860 levels from the 16D200 section of the privy, and the 16D203 drain samples generally contained between 100-200 opg (Table 5.2), which according to the Jones index has little significance for our interpretations. Pollen grains and *Ascaris* ova were also present in these samples. Recording the presence of pollen grains and what are likely roundworm ova is important as they indicate the preservation of other organic remains in the levels. The presence of *Pinus* or pine pollen grains is not surprising, as they are well adapted for transport of over a kilometre (Reinhard et al. 1986:34).

Four of the five fill layers from the 16C26 area of the privy dating to around 1870, contained between 100-300 opg, while the fifth layer contained none (Table 5.2). This would then indicate the presence of faeces as background noise in these levels, which is not necessarily an indication of intentional deposition. Pollen and *Ascaris* were also represented consistently in the counts. It should be noted that 16C65 contained over 20 *Ascaris* ova in both the first and second counts.

The deposits from the 16D25 privy, the 16C26 fill, the 16D203 drain, and concurrent 16D200 deposits, appear to indicate a very slight presence of faecal remains (Table 5.2). The three counts from 16D25, all of which contain parasite eggs, do suggest the use of this structure as a repository for human wastes amongst other materials.

Around 1875 a section of the 16D200 area of the privy appeared to have been filled very rapidly (Bouchard 1997:36). Parasite samples, 16D17, and 16D28 were processed from this episode and only 100 opg were noted for a total of six counts (Table 5.2). Layers from this event contained a high density of artifacts and very little soil, therefore the lack of parasite evidence would not be inconsistent with an interpretation of rapid trash fill from the surrounding yard and buildings. However, both levels did contain pollen grains and *Ascaris* ova, and 16D17 was very rich in raspberries, with the first and second counts each containing over 200 seeds.

Depressions over the 16D200 and the 16C26 sections of the privy, were filled during the late nineteenth century. Neither of these areas produced *Trichuris* eggs (Table 5.2), however, the second and third counts from the 16C18 samples contained evidence of *Pinus* or pine pollen, *Ascaris* or roundworm ova and what appeared to be the pollen of ragweed. This pollen was also found in the 16D18 sample and is perhaps pollen rain, like the pine pollen, from the surrounding environment.

The presence of abundant *Rubus* or raspberry seeds are considered indicators of faecal matter, and are the discerning factor in selecting which layers to process for parasite remains. They were therefore counted for during the processing of every parasite sample. The numbers of raspberry seeds within the 3 g samples taken for the parasite analysis (Table 5.2) are not necessarily consistent with the presence of ova, with the exception of 16D25, indicating that perhaps other factors must be considered in interpreting the trichurid fauna. For example, a single parasite sample from 16D38, in the drain, was processed. It contained no parasite ova, though 80 seeds were present, along with other seeds and pollen grains. And, no seed remains were recovered from the 16C72 samples, though the first and second counts yielded 300 and 100 opg respectively, along with pollen grains and what appeared to be *Ascaris* ova in every count. Finally, the 16D17 samples all contained *Ascaris* ova and pollen grains, while the second count yielded only 100 opg for *Trichuris*. This level contained the highest number of raspberry seeds, which number 200 seeds in the first and second counts, and five in the third count. Furthermore, this sample was thought to be depression fill, and it was not anticipated that any parasite ova would be recovered. Other factors, which will be considered in the following section, may have inhibited the preservation of the *Trichuris* ova. At this time, it is unclear whether the *Rubus* seeds are reliable indicators of faecal remains, a suggestion which will be discussed in the following section.

A number of factors which include the sampling of a privy structure, the ammonia-like odour of the soil encountered during both excavation and sample processing, the high numbers of raspberry seeds, and the overall excellent preservation of organic remains, suggest that there should have been significant levels of *Trichuris* encountered in this investigation. The results indicate the opposite and a number of plausible explanations must be considered.

Explaining the absence of *Trichuris*

There are several explanations that may account for the absence of abundant *Trichuris* ova in the Îlot Hunt deposits. These include poor egg preservation, lack of use for the Îlot Hunt levels for the deposition of faecal remains, consumption of anti-helminthic foods by the Îlot Hunt populations to combat parasitism, and methodological issues amongst others.

One simple explanation for the lack of trichurid abundance is that the eggs were missed in sampling. The basis for sample selection was the abundance of *Rubus* or raspberry seeds in the dry heavy fractions of the archaeoentomological samples. This is the standard means by which faecal samples are normally isolated in these studies (Fortin 1996:4; Reinhard 1992:240; 1994:63). Parallel sets of archaeoparasitological samples were taken during the excavations, though these samples were not rich in faecal remains, as had been hoped. As stated in the previous section, *Rubus* presence may not have been a sound criterion for sample selection. If this is so, then we must examine why *Rubus* is always considered a reliable faecal indicator. If it is not, due to a dietary preference or other factors, then some other means of isolating the faecal content of samples must be examined. Factors such as the concentration of *Rubus* seeds by surface water in the yard flowing into drains, and the presence of this species in urban yard sites where it favours growing on disturbed soils has generally not been considered in the archaeoparasitological literature, though deserves some treatment or experimentation.

The question of parasite egg preservation and decay must be considered. The addition of lime to the privy, possible *in situ* destruction by fungus or bacteria, and the use of washing soda in sample processing may have resulted in the preservation of fewer *Trichuris* eggs. Sample 16K5 was processed without washing soda and the first count from this sample contained 200 opg, while the second and third counts contained *Ascaris* but not trichurid remains. These findings are consistent with other samples processed from the site, so the use of washing soda does not appear to be a factor which affected egg preservation.

Lime may have been added to the privy destroying some of the parasite ova. A student at Université Laval, Martine Duchaîne, treated a column of samples from the 16D200 area of the privy to chemical analysis. She found high levels of calcium carbonate ($CaCO_3$) (Martine Duchaîne, personal communication 1999). Lime is primarily made up of calcium carbonate and is used to make plaster and mortar. Lime can also be used as a fertiliser to neutralise acidity, and can be highly caustic. In the historic period, lime was, at times, thrown in privies as a disinfectant. The presence of the calcium carbonate may also be a result of the presence of shells or plaster, both commonly found in many of the deposition layers. Evidence for the use of lime in Québec's privies is mentioned in "Instructions générales et personelles aux familles et individus" dating to 1854, issued by the Central Health Office of Québec. Due to their fear of the coming cholera epidemic, they issued general sanitary guidelines which included in Article 9, "que toutes caves, grilles, étangs, commodités, et places contenant des matières ou substances malsaines… soient immédiatement nettoyés, vuidés, changés ou réparés et bien arrosés de chaux" (Bureau central de santé 1854:9). Another example of liming of privies is from New York City, where, in 1802, the Board of Health recommended that a work force be created that disinfected the city's privies with lime (Geismar 1993:60). Geismar goes on to suggest that perhaps the addition of lime would have created an enriched soil, rather than an organic deposit, inhibiting the preservation of parasite eggs (Geismar 1993:65). She also goes on to point out that if a vacuum system was used to clean out the privy, it may have eliminated human excreta, but left large debris still in the privy (Geismar 1993:65). The addition of lime to the Îlot Hunt layers may have affected the preservation of the parasite ova.

Fungus or bacteria may have been present in these privy layers and destroyed the ova. This has also been suggested as a possible cause for a lack of some ova in a Boston privy (Driscoll 1995:6). *Ascaris* ova are virtually impenetrable, while *Trichuris* ova, contain lower levels of chitin (also found in the exoskeletons of Coleoptera, ensuring their preservation), and are made up of mostly protein. It has been suggested that while *Ascaris* are resistant to decay, trichurids may be susceptible to fungal penetration (Jim Smith, personal communication 1999). This explanation would then account for the preservation of the *Ascaris* ova, while *Trichuris* was reduced to a very small fauna.

Some treatments, such as the consumption of figs, and the taking of *Chenopodium* oil are known to have anti-helminthic properties (Chandler 1955:395), as is *Daucus carota* or Queen Anne's lace (Driscoll 1994:33), and peaches (Driscoll 1995:12). *Ficus* or figs and a peach pit were found in the palaeobotanical remains from the 11C33 privy sample (Fortin 1996:17), and were also noted during the processing of the archaeoentomological samples from the privy. *Chenopodium album* was found in levels 11B46 and 11B51 (Fortin 1996:17), of the 16C26 area of the privy. Archaeologists believe that in New York City's nineteenth-century Five Points district the Irish population likely controlled *Ascaris* with the oil of *Chenopodium* made from *C. ambrosioides* (Yamin 1997:50). The Five Points site archaeobotanist identified seeds of chenopodia from soil samples confirming its presence. Known today as Mexican-Tea or American wormseed, this is an introduced member of the goosefoot family which grows on waste ground. At least three different species of *Chenopodium* can grow in the Québec City region (Peterson and McKenny 1968:378), and several species in this genus appeared to work as vermifuges, to expel worms.

Nineteenth-century recipes from four Canadian 'home recipe books' (Anonymous 1882, 1895; Chase 1864; Moore 1874) indicate the prevalence of worms and their treatments. Chase's volume contains four recipes for vermifuges and includes graphic testimonials as to their effectiveness in expunging worms, which include roundworm or *Ascaris lumbricoides* (Chase 1864:164-170). Ingredients for vermifuges, 'worm cakes' and 'worm teas' include combinations of chenopodii or wormseed (*Chenopodium*), pumpkinseed, turpentine, oil, clove oil, elm bark, garlic, salt, castor oil, valerian, peppermint, capsicum, alum, pomegranate bark, sugar, and gum tragacanth. One of these volumes highlights the medical qualities of pumpkins, which the text states are a good ""antidote against the animaculae which infest the bowel", which is likely a reference to parasites. The text recommended making a paste of pumpkin seeds and castor oil to eject any such beings (Anonymous 1882:286).

James Langstaff's nineteenth century medical diary (Duffin 1993) states that, especially for children, worms were a constant problem. His records refer to 'worms', 'ascarides', and 'taenia', as well as scabies and lice. He used vermifuges, powders, lozenges and lotions to treat these conditions (Duffin 1993:110).

Due to the prevalence of worms in this historic period, indicated by the evidence for worm recipes, and Langstaff's records, parasites were common afflictions, and most likely affected the Îlot Hunt site population. Due to the urban density of the Lower Town area, the inhabitants of the site would have had to purchase their fresh produce from local farmers who may have used excrement as fertiliser. Both *Chenopodium* or wormseed and pumpkin seeds were found in the privy, though their presence in no way proves they were consumed as vermifuges.

A final explanation for the lack of *Trichuris* worth examining, is their absence due to lack of deposition. The Îlot Hunt site was located, in part, on one of the important streets in Lower Town, rue Saint-Pierre. This street contained many of the important businesses in the city, though their proprietors mostly lived in the Upper Town according to the city directories. Perhaps the importance of this area resulted in sewer installation during the earliest phases of construction of the city-wide system. If this were the case, at the very least, the Hunt House and the adjacent buildings directly south of it would have had an alternative means of waste disposal for human wastes as early as the 1860s, and the already constructed 16D200 privy structure could have been used for general trash, as evidenced by its contents. Unfortunately, no documentation to test this hypothesis was available, though it would account for at least part of the 'missing' human faeces. As the site, throughout the entire period under consideration, was made up of several dwellings of mixed socio-economic incomes and commercial uses, it was always assumed that there would have been some faeces in the levels examined. However, if, as the insect faunas seem to indicate, the 16C26 area of the privy was not used primarily for domestic wastes and instead for commercial and stable wastes, and if sanitary services were available for the dwellings contributing to the domestic trash deposits in the 16D200 levels, then perhaps the assumption of high faecal contents in the privy was unfounded.

Several explanations have been considered to account for the absence of abundant *Trichuris* ova from the Îlot Hunt deposits. Other archaeoparasitological studies based on excavations of historic privies (Driscoll 1995; Fries et al. 1990; Reinhard et al. 1986), contained substantially higher levels of parasites than found at Îlot Hunt. Of the explanations considered, it seems that the most likely cause of the poor parasite egg recovery was due to either the destruction of *Trichuris* eggs after deposition or low faecal concentrations in the Îlot Hunt samples analysed.

6.4 Nineteenth century health and sanitation at Îlot Hunt from an environmental, textual and material perspective

Around 1850, the Îlot Hunt site working and inhabiting population was 38 people from at least four extended families (Bureau d'enregistrement et de statistiques 1853-1856). The majority of the occupants were Irish, and many were unskilled labourers, who likely settled in Québec City due to the previous two decades of heavy immigration. Businesses on the site included commission agents, a general merchant, two general brokers, two sailmakers, and a tavern. This densely occupied parcel of land contained four houses on rue Saint-Pierre and rue Saint-Antoine, three warehouses, a bakery, two stores, and stables amongst other secondary buildings (Leclerc 1998:27). The structures on the site were not greatly altered until 1875.

Sanitary services on rue Saint-Pierre and rue Saint-Antoine included drains along the street edges similar to those found throughout the city at this time. The site itself contained two privies, two drains, and a former cistern or well which was filled around 1850. The 16D25 privy, which appears to have been a covered wooden structure, received some human wastes as well as trash from at least one of the homes on site. The 16C26 section of the 16D200/16C26 privy contained wood detritus, sawdust, and domestic trash which may have included floor sweepings. Some water was in or around this deposit.

City inhabitants experienced continuing cholera epidemics, interspersed with bouts of typhus and influenza. An attempt at disease prevention may account for the infilling of the well or cistern structure, 16E100, as stagnant sources of water, according to the medical beliefs of the time period were considered a danger (Goulet and Keel 1993:419). It did appear to contain water at the time of its infilling, evidenced by the downward percolation of the fill contents. Some domestic trash was thrown into this structure, but the majority of discarded debris appears to have been yard soil and debris used to fill in this structure.

The few material culture remains indicating sanitary or medical practices from this period (Table 3.2) include a chamber pot, a glass vial, a medicinal bottle, a pitcher, and a bottle which contained shoe blackening or polish. This last artifact may have been used by one of the professionals or merchants. The artifacts recovered indicate a mixed population of lower and middle income occupants, who prescribed to typical nineteenth century medical treatments such as the use of patent medicines, while their homes contained the usual household visitors of roaches, bedbugs and carpet beetles.

At some point during the next ten-year period, the 16D25 privy was filled in and abandoned, and the 16D200/16C26 privy structure was cleaned out. The drain running into the 16D200 structure was filled with debris, and deposits built up again in both sections of the large privy structure, which may have still had a functioning drain at its eastern end. However, it is unclear what this would have drained into, as the area east of the privy contained buildings.

At least 20 people lived or worked on the site around 1860 (Census Department 1863-1864). Businesses included wine and flour merchants amongst others, a tavern, a sailmaker, insurance agents, a hardware store, and a jeweller. As the site was located in an economically important part of the city, infrastructure services may have become available to the site's occupants during the decade between 1860-1870. This availability could have provided the impetus to fill and abandon the 16D203 drain, and to discard the many chamber pots found within it, if indoor plumbing was now available.

Another explanation for the drain's discontinuation is found in the small mammal component of the archaeozoological remains. Evidence of several rats was found in the drain (Boucher, personal communication 1999), providing a plausible reason for its infilling. If the drain was connected to the Hunt House, or one of the structures between the privy and the house, such as the building dating to c. 1815 (Appendix A, Figure 18), the access of the thriving rodent fauna to this structure via the drain, may have inspired its abandonment. Other contents in the drain included kitchen wastes and some wood. The contemporaneous16D200 deposits indicate the site's inhabitants were filling the privy with all sorts of household and yard debris, including mammal excrement and wood. The palaeobotanical remains concur that this was a typical latrine assemblage (Fortin 1996:11-12). The 16C26 area of the privy contained sawdust and less domestic trash, while the majority of its insect fauna was derived from the yard and the commercial buildings.

Lice were present on site and may also be indicated by the recovery of a fine-toothed comb. The brushes, vials, ointment jars and medicinal bottles, as well as the recovery of Spanish Fly fragments, indicate the continued use of patent medicines and treatments.

In 1866, again in fear of an oncoming cholera epidemic, a municipal sanitary by-law was passed and many privies were cleaned out (Dagneau et al. 1983:156). Whether by the site's users, or city workers, the 16D200/16C26 privy was again subjected to another cleaning episode before the deposition of the 1870 and 1875 fills in privy sections 16C26 and 16D200 respectively.

Québec City suffered a recession beginning around 1870. The site's population of at least three families, one of which had a servant, indicates an increase in population from the previous decade. The commercial dwellings remained fairly constant during this period, though a warehouse was added to the site at this time (Leclerc 1998:28). The faunal analysis does, however, indicate that fewer animals were being raised on site, perhaps an indication of economic stress (Boucher 1999:102) or due to respect for the city by-laws limiting the stabling of animals in the city.

The 16C26 section of the privy dating to around 1870 was filled with bones, animal wastes, sawdust, wood, and some domestic trash. The large volume of bones recovered from these levels may have been due to the disposal of trash from the tavern or hotel on rue Saint-Antoine. The artifacts related to personal health and sanitation indicate the continued use of patent medicines and Spanish Fly for medical treatments, while the appearance of a toothbrush in this level is interesting as it may be considered a status indicator of middle class homes (Howson 1993:154). As one of the families living on site was able to afford a servant (Department of Agriculture 1873-1878), its presence is not surprising.

In 1875, the site underwent large structural changes as rue Dalhousie was extended which literally cut through three warehouses on the site dividing the site into two parts (Leclerc 1998:29, Appendix A, Figure 16). The last deposits from within the 16D200/16C26 privy structure date to this time period. The on site changes provided the impetus to fill in and abandon the privy structure, as a building whose foundation was found during the 1993 and 1995 excavations, was erected over the southern part of the privy. These final privy fill deposits from the 16D200 section contain little soil and a great density of artifacts, glass, metal, and bones. The insect fauna indicates some household and yard trash, but lacks many of the groups seen in lower privy levels, such as the outdoor, phytophagous, carrion, and synanthropic faunas. This seems to indicate that these levels were filled specifically with household artifacts and yard debris, as beetles found in floor sweepings and those representing the surrounding yard did not make their way into this deposit.

Included in these levels were chamber pots, vials, bottles, and ointment jars, indicating the ongoing use of traditional medical treatments, further confirmed by the presence of fragments of Spanish Fly. The presence of 10 perfume bottles, four pitchers and five shoe polish bottles, appear to indicate attention to personal care by occupants of the site. Bouchard found, in his analysis of ceramics from c. 1875-1900, that, during this time period, the occupants of the site had greater access to goods of higher value than those in Saint-Roch, a working class sector of the city (Bouchard 1998:91), perhaps accounting for the disposal of these artifacts.

The site contained varied commercial businesses during the last two decades of the nineteenth century, though there is a rise in the number of artisans on site (Bouchard 1998:32). The population of the site contained five families, one of which had a servant. The site's population peaks around 1891, with 39 people and seven families, and then declines slightly over the next ten years (Department of Agriculture 1882-1885, 1893-1897). At this time, between 1880-1900, water and sewer services may have been available for the rue Saint-Antoine buildings (Simoneau and Rouleau 1988:89), though the installation of these structures is not reflected in the beetle or artifactual faunas.

As the port industries declined, and less commercial space was needed, it appears that the users of the site adapted. They did so by renovating buildings which included some of the commercial structures and the c. 1824 house on rue Saint-Antoine, creating two dwellings.

Two deposits represent the infilling of levels due to the compaction of the organic privy fills, while a virtually sterile fauna, due to the abundance of limestone in the study sample, was associated with the renovation of the 1875 building. The deposit over the 16D200 section of the privy contained dryish fill with sand, wood, animal wastes, and domestic trash. The insects recovered in the depression over the 16C26 fill indicated a wetter deposit, with less kitchen wastes, though they did contain animal wastes and evidence of the continued use of Spanish Fly.

The late century artifacts documenting personal care indicate households with ready access to material goods as they discarded three combs, three pitchers, and a toothbrush. Otherwise, chamber pots, and medical bottles recovered do not detail any changes in the occupants' approaches to medical applications.

If the site's occupants were wealthier, as indicated by their ceramics, and some of the artifacts related to personal hygiene, they may have been at this time, employing a service for waste removal. The years between 1880-1900 indicate a densely occupied site, without any sanitary structures available for disposal of yard and domestic trash. Even if they were connected to the city sewer and water services, other debris from the homes, commercial establishments and artisans still needed disposal. Though the city's official dump was at Pointe-aux-Lièvres, the proximity of the river may have provided a nearby location to discard any unwanted trash once the depressions over the 16C26 and 16D200 privy sections were filled.

The Îlot Hunt site occupants during the fifty-year period under consideration, were always a mixture of site users and inhabitants. They represented a broad spectrum of ethnicities, occupations and income levels. It appears that they took care during this time period to maintain some order in the yard. The density of use of this urban space necessitated at least two cleaning episodes of the large privy structure between 1850-1875. Three other structures, 16D203, 16E100 and 16D25 were also filled at this time. These events may reflect Québec City's battle with epidemic diseases, as the site's occupants filled in structures that they perceived to be dangerous to their health. It is assumed that city sewer and water services were available for the latter half of this study period, and that disposal of unwanted wastes, other than those that filled the two yard depressions, were removed off the site. The period from 1875-1900 saw an unprecedented density of site users as another building was added to already the crowded yard, and the c. 1824 house on Saint-Antoine was subdivided. While sanitation patterns did indeed change over this fifty-year period, personal hygienic habits and the use of traditional medical treatments remained the same.

6.5 What new evidence can this study shed on the nineteenth century?

This multi-disciplinary study of medical and sanitary practices at the Îlot Hunt site is valuable as it indicates our perceived notions of the past may be inaccurate if we only consider a single data source upon which we base our interpretations. If the CHPQ inspectors' reports, the primary narrative texts available to us which describe Lower Town at this time (ANQ, Rapports du CHPQ, 1888, 1889, 1891, 1893, 1898), were our only tool with which to study Îlot Hunt, then the site would have been considered filthy and almost unfit to live in. These reports were written in the last two decades of the nineteenth century, and cast the city in a particularly bad light. The occupants of the densely populated Îlot Hunt site, 40 years previous to the CHPQ reports, appear to have maintained the various structures of their yard area, and cleared them of their contents occasionally

This disparity seen between the documentary and archaeological and environmental data is similar to that found in Yamin's study of New York's Five Points neighbourhood (Yamin 1997, 1998). Documentary evidence describing that site portrayed it as a notoriously unsavoury area, rife with gambling and prostitution. In her analysis of data from a variety of sources, Yamin was able to indicate that many of the site's inhabitants, members of New York's thriving immigrant communities, maintained respectable homes (Yamin 1997:48). She also points out that middle-class rhetoric describing nineteenth-century working-class districts throughout the English-speaking world used the same vocabulary (Yamin 1997:52), which elaborated on the social and physical ills of these areas.

In this Îlot Hunt study, we must therefore reconsider the CHPQ reports (ANQ, Rapports du CHPQ, 1888, 1889, 1891, 1893, 1898), and question whether they are representative of Québec City's Lower Town. Parts of Lower Town, such as the Saint-Roch ward, were primarily working class, though the Îlot Hunt site always maintained a population of both the working and middle classes. Unfortunately, we do not know if the reports' authors recognised these various components of the Lower Town. Furthermore, Joseph Beaudry and Elzéar Pelletier, the CHPQ inspectors, were qualified medical doctors. Such practitioners, allied with other members of the middle class, made up the greatest advocates of the hygienic reform movements of North America at this time, and would likely present a portrait of Québec City from the perspective of a committed sanitary reformer. Such a position may have been overtly critical of ongoing sanitary practices that did not concur with their views. Finally, we must also consider whom the reports were written for. Perhaps their intended use was to push the municipal authorities of Québec City to act in the best sanitary interests of their citizens, and to fund necessary changes. If this was the case, the inspectors may have intentionally exaggerated the plight of the Lower Town as such scare mongering had been found to be an effective political tool. While the intentions and mindset of the authors of these and any other historical narratives can never be fully known, it is worthwhile to consider alternative explanations for their content. This is particularly useful in cases such as this one where the texts do not accurately describe the late nineteenth-century Lower Town, at least not at the Îlot Hunt site. For historical archaeologists, such findings provide a valuable lesson in the critical and judicious use of textual data.

The insects recovered from the site also indicate that these nineteenth-century structures were not as infested as anticipated. While some degree of infestation was experienced, the amount demonstrated by the identified insect assemblages by no means approached what could be classified as a serious infestation, though the samples processed only provide a partial picture of the past.

This study was also able to shed light on medical practices of the nineteenth century. The use of Spanish Fly appears to have been extensive if we consider its consistent presence in layers between 1860-1900. As no evidence of its use was indicated on the mostly plain medicinal bottles and vials, if they were in fact the containers in which it was procured, confirmation of its presence in the archaeoentomological assemblage is the sole means to prove its use at the site. This appears to be the first discovery of *Lytta vesicatoria* from a North American archaeological site. With knowledge of its presence on this site, we may speculate as to what ailments plagued our site's inhabitants and further understand people'sperceptions of illnesses and treatments in the nineteenth century.

This project has investigated nineteenth-century levels of infestation, hygiene, sanitation, and medical treatments. It has indicated that the content of nineteenth-century textual sources must be treated with some caution, preferably in conjunction with other remains such as artifacts, structural remains, and environmental data. The multidisciplinary approach applied to this Îlot Hunt study contributes to our understanding of the site, its people and their practices, by its critical evaluation and consideration of the site's many contexts

.

CHAPTER SEVEN

CONCLUSIONS

7.1- RESTATEMENT AND CONSIDERATION OF HYPOTHESIS

For this Îlot Hunt site study, our primary hypothesis examined the *promotional perspective* of the period between 1850-1900. This states that if we consider medical advances, the development of urban infrastructures, and the creation of municipal, provincial and federal by-laws, then an overall improvement in late nineteenth-century life was experienced. On the Îlot Hunt site we anticipated a declining rate of parasitism, the abandoning of privy pits, and the installation of indoor plumbing or at the very least, the existing structures would be connected to the city's water and sewer networks. We should also be able to document declining infestations in houses and foodstuffs, and evidence of fewer patent medicinal remedies represented by medical bottles, as new scientific principles filter into everyday medical practices. The yard environment should be generally cleaner, as people take measures to ensure a more sanitary urban environment.

Our null hypothesis states that, considering the *reported perspective*, new hygienic standards are not being integrated into people's lives and existing sanitary systems were not improved. Parasitism in the population would therefore remain at a constant between 1850-1900, and there will be the continued use of traditional medical remedies. The yard will appear consistently unclean throughout this period, and the homes and commercial dwellings of the site will continue to suffer insect infestations due to their poor domestic hygiene, further evidenced by the ongoing use of privies, even though new sanitary facilities were available.

Between the 1850-1900, we witness the end of the use of privy pits at the Îlot Hunt site, the infilling of drains and other structures, and what appears to be the connection to city sewer and water networks. There is no evidence on this site of the high degree of parasitism anticipated in the site population. This may be the result of post-depositional processes destroying the parasite eggs, or, a lack of human faeces in the samples, as the site's users were connected to city services sooner than anticipated. While some infestation by ectoparasites and grain and stored products was experienced by the site's users, the presence of these species is not unexpected for a nineteenth century site, and their numbers do not constitute an infestation beyond what we would expect for normal household levels. As depressions formed due to compaction of lower privy levels, the surface hollows were filled with unwanted trash.

Documentary data from late nineteenth century reports of Québec City's Health Office and the Provincial Hygiene Inspectors' reports lament the ongoing use of privy structures, and generally poor sanitation throughout the city, and particularly in Lower Town, where the Îlot Hunt site is located. Our site does not conform to this. The location and population of Îlot Hunt, which at its western edge was situated on one of the most important streets in the city, may have been a factor in the resulting sanitation experienced in the yard area. The occupants of the site, throughout the entire period, were a mixture of labourers, professionals, artisans, and families. Their access to better material goods improved between 1850-1900, even though the city experienced a recession beginning around 1870. If the Îlot Hunt site was entirely populated by working class families in a different sector of the city, perhaps they would not have had the same access to services. Instead, on this site, the more middle class occupants of the site may have had the means to ensure better sanitary services.

While we are unable to accurately comment on the level of parasitism experienced by the site's occupants, artifacts such as combs, perfume bottles and brushes indicate ongoing attention to personal hygiene. The discovery of Spanish Fly in levels dating to 1860, 1870, 1875, and 1880-1900, along with glass vials and medical bottles, represents an unchanging adherence to traditional medical practices, though it seems that good sanitary practices were in place at this site.

This human reaction at the Îlot Hunt site, to the changing nature of the Lower town in the nineteenth century, was also examined by Goyette (1999) in her analysis of the 1993 Îlot Hunt excavation results. Goyette's results, as well as those found here, provided similar interpretations which documented the inhabitants' systemic adaptation to their situation demanded by the local economy, with the yard's material and structural contents representing their sanitary and economic practices (Goyette 1999:164-165).

Our primary hypothesis, therefore best reflects the reality experienced by Îlot Hunt site occupants at this time. While medical practices did not change during this period according to the evidence available, and the level of parasitism is not available, the occupants did appear to conform somewhat to the city by-laws which demanded the occasional cleaning of privy structures, and infilling of areas potentially harmful to their health. The low levels of infestations in the homes and dwellings around the site confirm the regular clearing and removal of household and yard wastes. While the documentary data, as suggested by our *reported perspective*, for the city may indicate another reality, the Îlot Hunt site does not conform to it.

7.2 CONCLUDING COMMENTS ON THE USE OF THE ÎLOT HUNT METHODOLOGY

This study endeavoured to combine data from a number of sources to elucidate information about sanitary and medical practices during the latter half of the nineteenth century at the Îlot Hunt site. These sources included documents, material culture, and environmental data, primarily from preserved insect and parasite fauna, though archaeobotanical and archaeozoological results were also integrated into this study. The data collected for this project varied widely in both quantity and quality, though these factors were not necessarily in our control.

Many documents referred to the sanitary situation in Québec City at this time, providing an overall view of the city. These sources include documents from the municipal and Provincial Boards of Health, city by-laws, committees, and inspectors' reports. While the personal accounts of the city are subjective and must be treated as such, they mention invaluable statistics and comment on this area of the city. Documentation referring to sanitary infrastructures on the streets adjacent to our study site could not be located for the years between 1850-1900, and it was hoped the material remains would account for their absence.

Material culture analysed in this study included structures and artifacts related to sanitation. Structural remains in the yard area existed for the first half of the time period under consideration, while excavations during the 1980s on rue Saint-Antoine indicated traces of infrastructure that may also fit into our time period.

When the artifacts related to hygiene and sanitation are considered in comparison with other contemporaneous sites located in Northeastern North America (Bond 1989; Garrow 1996; Howson 1993), the Îlot Hunt assemblage, totalling only 74 artifacts, does not contain abundant indicators with which we can trace personal health and sanitation. These artifact volumes, considering the number of users of the site throughout this fifty-year period, suggest the disposal of remains in other locales such as the river or in the city dump.

The examination of soil samples from this site did not yield the anticipated parasite results, due to the use of the sewer service throughout most of the time period under consideration here, or due to the post-deposition destruction in situ of parasite ova. The lack of *Trichuris* ova in the examined archaeoparasitological samples has been examined, and one surprising factor was the lack of parasite remains when the amount of *Rubus* seeds are considered. The use of these seeds as a reliable, consistent, faecal indicator must be questioned. Unfortunately, we have no way of proving this, though further testing of this is suggested. The insect fauna, however, derived primarily from samples taken systematically during the 1995 excavations, yielded a rich and varied fauna totalling 6755 individuals. Only a percentage of this large fauna provides specific environmental and cultural data aiding interpretation of the site's many contexts. As the largest insect fauna studied from a North American archaeological site to date, this assemblage is important, as it will provide a comparative framework for future urban archaeoentomological studies. The palaeobotanical and archaeozoological analyses undertaken by Fortin (1996) and Boucher (1999) were extremely useful in their confirmation or further questioning of the results of the insect and parasite analysis. The Îlot Hunt study, with several parallel means with which to examine the site's many contexts, is an excellent model of the integration of biological data into a contextual archaeological study.

Another important aspect of this study is realisation of the dominance of the nineteenth-century urban fauna by introduced species. A full 79 percent of the insects that were identified to the species level were introduced or non-native species. Many of these are some of the most serious pests of stored products, though their numbers in the Îlot Hunt levels do not indicate any infestation beyond what would have been normal for this time period. Archaeoentomology can provide valuable chronological information, supplementing our knowledge of histories of introduced species in North America that would otherwise be unobtainable. Such data may also give rise to accurate models concerning rates of insect adaptation and dispersal across new landscapes, theories that are otherwise left to supposition.

While each of these data sources are limited in their abilities to answer specific archaeological questions, their combined results provide sensitive data about many of the site's contexts. Only by the judicious integration of such multiple data sets in historical archaeology can we ever hope to gain greater access to subject areas such as health and sanitation.

REFERENCES CITED

Addyman, P.V.
1982 The archaeologist's desiderata. In *Environmental archaeology in the urban context*, edited by Allan R. Hall and Harry K. Kenward, pp: 1-5. Research Report No 43, The Council for British Archaeology, London.

Agriculture Canada
1902-1906 *Fourth Census of Canada, 1900-1901*. S. E. Dawson, Ottawa.

Aitken, A.D.
1975 Insect Travellers. Volume 1. Coleoptera, *Technical Bulletin of the Ministry of Agriculture, Fish and Food* 31, England.

Allison, M.J., A. Pezzia, I. Hasegawa, and E. Gerszten
1974. A Case of Hookworm Infestation in a pre-Columbian American. *American Journal of Physical Anthropology* 41: 103-106.

Anderson, R.S. and S.B. Peck
1985 The Carrion Beetles of Canada and Alaska (Coleoptera: Silphidae and Agyrtidae). *The Insects and Arachnids of Canada* 13. Biosystematics Research Institute, Ottawa.

Anonymous
1882 *Home and health and compendium of useful knowledge: a cyclopedia of facts and hints for all departments of home life, health, domestic economy*. The Advertiser Printing and Publishing Co., London, Ontario.
1895 *Common sense recipe book containing all the latest recipes...and also very valuable medical recipes*. John Lovell & Sons, Montréal.

Araújo, A.J.G.
1982 Oxyurid (Nematoda) Egg from Coprolites from Brazil. *Journal of Parasitology* 68(3): 511-512.

Atkinson, T.C., K.R. Briffa, and G.R. Coope
1987 Seasonal temperatures in Britain during the past 22,000 years, reconstructed using beetle remains. *Nature* 325: 587-592.

Bain, A.
1997a *Analyse des restes archéoentomologiques de l'îlot Hunt (CeEt-110), dans la Basse-ville de Québec*. Submitted to CELAT, Université Laval, Québec, Québec.
1997b *Analyses des restes archéoentomologiques du site Faubourg (BjFj 56-95), Montréal, Québec*. Submitted to Arkéos, Montréal, Québec.
1997c *Archaeoentomological Analysis at the Ferryland Site, Newfoundland*. Submitted to Archaeology Unit, Memorial University of Newfoundland, St. John's Newfoundland.
1998a *Analysis of Insect Remains from the Abiel Smith School Site, Boston, Massachusetts*. Submitted to The Museum of Afro American History, Boston, Massachusetts and United States National Park Service, Lowell, Massachusetts.
1998b A Seventeenth Century Beetle Fauna from Colonial Boston. *Historical Archaeology* 32(3): 38-48.

Bain, A. and L. LeSage
1998 A Late Seventeenth Century Occurrence of *Phyllotreta striolata* (Coleoptera: Chrysomelidae) from North America. *Canadian Entomologist* 130: 715-719.

Baugher-Perlin, S.
1982 Analysing Glass Bottles for Chronology, Function and Trade Networks. In *Archaeology of Urban America: The Search for Pattern and Progress*, edited by Roy S. Dickens Jr., pp: 259-290. Academic Press, New York.

Beaudry, M.C.
1989 Introduction. In *Interdisciplinary Investigations of the Boott Mills, Lowell, Massachusetts, Volume III, The Boarding House System as a Way of Life*, edited by Mary C. Beaudry and Stephen A. Mrozowski, pp: 1-5. Cultural Resources Management Studies No. 21. U.S. Department of the Interior, National Park Service, North Atlantic Regional Office, Boston, Massachusetts.
1993 Public Aesthetics versus Personal Experience: Worker Health and Well-being in 19th-Century Lowell, Massachusetts. *Historical Archaeology* 27(2): 90-105.

Beaudry, M.C. and S.A. Mrozowski (editors)
1989 *Interdisciplinary Investigations of the Boott Mills, Lowell, Massachusetts, Volume III, The Boarding House System as a Way of Life*. Cultural Resources Management Studies No. 21. U.S. Department of the Interior, National Park Service, North Atlantic Regional Office, Boston, Massachusetts.

Beer, R.J.S.
1976 The Relationship between *Trichuris trichiura* (Linnaeus 1758) of man and *Trichuris suis* (Schrank 1788) of the pig. *Research in Veterinary Science* 20: 47-54.

Bernier, J.
1989 *La médecine au Québec: Naissance et Évolution d'une profession.* Les Presses de l'Université Laval, Québec.

Bervin, G.
1991 Québec au XIXe siècle. Les éditions du Septentrion, Sillery, Québec.

Bilson, G.
1980 *A Darkened House: Cholera in Nineteenth Century Canada.* University of Toronto Press, Toronto.

Blatchley, W.S. and Leng, C.W.
1916 *Rhynchophora or Weevils of North Eastern America.* The Nature Publishing Company, Indianapolis.

Bond, K.H.
1989 The medicine, alcohol, and soda vessels from the Boott Mills Boardinghouses. In *Interdisciplinary Investigations of the Boott Mills, Lowell, Massachusetts, Volume III, The Boarding House System as a Way of Life* edited by Mary C. Beaudry and Stephen A. Mrozowski, pp: 121-139. Cultural Resources Management Studies No. 21. U.S. Department of the Interior, National Park Service, North Atlantic Regional Office, Boston, Massachusetts.

Boroweic, L.
1987 The genera of seed beetles (Coleoptera, Bruchidae). *Polskie Pismo Entomologiczne* 57(1): 3-207.

Borror, D.J. and R.E. White
1970 *Les Insectes de l'Amérique du Nord (au nord du Mexique).* Houghton Mifflin, Boston.

Bouchard, P.
1997 *Le site de l'îlot Hunt à Quebec (CeEt-100) : rapport annuel d'activités, cinquième campagne de fouilles archéologiques.* Submitted to Université Laval, Québec and Ville de Québec, Québec.
1998 *Le site de l'îlot Hunt à Québec (CeEt-110) : étude socio-économique des habitants d'après la collection archéologique, 1850-1990.* Unpublished M. A. thesis, Département d'histoire, Université Laval, Québec, Québec.

Boucher, G.
1999 *Les habitudes alimentaires des habitants de l'îlot Hunt (CeEt-110) de 1850 à 1900 : étude archéozoologique.* Unpublished M. A. thesis, Département d'histoire, Université Laval, Québec, Québec.

Bousquet, Y.
1990 *Beetles associated with stored products in Canada.* Research Branch, Agriculture Canada, Ottawa.
1991 *Checklist of Beetles of Canada and Alaska.* Minister of Supply and Services, Ottawa.

Bousquet, Y. and S. Laplante
1999a Les Coléoptères Histérides du Québec. *Fabreries* Supplément 8. Association des entomologistes amateurs du Québec, Varennes.
1999b Taxonomic review of the Canadian species of the genus *Monotoma* Herbst (Coleoptera: Monotomidae). *Proceedings of the Entomological Society of Ontario*, in press.

Brace, C.
1995 Public Works in the Canadian City; the Provision of Sewers in Toronto 1870-1913. *Urban History Review* 23(2): 33-43.

Breese, M.H. and T.E. Wise
1959 The biology of *Nausibius clavicornis* (Kug.) (Col., Cucujidae). *Bulletin of Entomological Research* 50: 237-258.

Bright, D.E.
1976 The Bark Beetles of Canada and Alaska (Coleoptera: Scolytidae). *The Insects and Arachnids of Canada* 2. Biosystematics Research Institute, Ottawa.

Brothwell, D.
1986 *The Bog Man and the Archaeology of People.* Harvard University Press, Cambridge, Massachusetts.

Brown, H.P.
1972 *Aquatic Dryopoid Beetles (Coleoptera) of the United States.* Biota of Freshwater Ecosystems, Identification Manual No. 6. Environmental Protection Agency, Washington.

Brown, W.J.
1940 A key to the species of Ptinidae occurring in dwellings and warehouses in Canada (Coleoptera). *Canadian Entomologist* 72: 115-122.

Buchanan, L.L.
1936 The *Pacificus* group of *Cossonus* (Coleoptera: Curculionidae). *The Pan-Pacific Entomologist* 12(3): 111-116.

Buckland, P.C.
1976a The Environmental Evidence from the Church Street Roman Sewer System. *The Archaeology of York, The Past Environment of York* 14/1. Council for British Archaeology for York Archaeological Trust, London.
1976b *Niptus hololeucus* (Fald.) (Col., Ptinidae) from Roman Deposits in York. *The Entomologist's Monthly Magazine* 111: 233-234.
1978 Cereal production, storage, and population: a caveat. In *The effect of man on the landscape: the Lowland Zone*, edited by Susan Limbrey and John G. Evans, pp: 43-45. Research Report 21, Council for British Archaeology, London.
1981 The Early Dispersal of Insect Pests of Stored Products as Indicated by Archaeological Assemblages. *Journal of Stored Product Research* 17(1): 1-12.
1982 The Malton Burnt Grain: A Cautionary Tale. *The Yorkshire Archaeological Journal* 54: 53-61.
1987 A Tale of a Flea. *Antenna* 11: 135-7.
1988 North Atlantic faunal connections - introductions or endemics? *Entomologica Scandinavia Supplementum* 32: 8-29.

Buckland, P.C., A.C. Ashworth, and D.W. Schwert
1995 By-passing Ellis island: Insect Immigration to North America. In *Ecological Relations in Historical Times*, edited by Robin Butlin, and Neil Roberts, pp: 226-244. Blackwell, Oxford.

Buckland, P.C., J.R.A. Greig, and H.K. Kenward
1974 York: an Early Medieval Site. *Antiquity* 48: 25-33.

Buckland, P.C., and J.P. Sadler
1989 A biogeography of the human flea, *Pulex irritans* L. (Siphonaptera: Pulicidae). *Journal of Biogeography* 16: 115-120.
Bureau central de santé
1854 Règlements & c. adopté par le Bureau central de santé sous l'acte 12 Vict. Cap. 8. Canada Gazette, Québec. Bureau d'enregistrement et de statistiques 1853-1855 *Rencensement des Canadas.* J Lovell, Québec.
Bureau de Santé, Canalisation de la ville
1849 Bureau de Santé, Canalisation de la ville, Rapport d'un Comité Spécial du Bureau de Santé sur un meilleur sytème de canalisation de la ville. Imprimerie du 'canadien' No. 9, Rue La Montagne, Basse-Ville, Québec.
Busvine, J.R.
1976 *Insects, Hygiene, and History*. Athlon Press of the University of London, London.
Butzer, K.
1980 Context in Archaeology: an Alternative Perspective. *Journal of Field Archaeology* 7: 417-422.
1982 *Archaeology as human ecology: Method and Theory for a contextual approach.* Cambridge University Press, New York.
Cambefort, Y.
1994 *Le Scarabée et les Dieux*. Boubée, Paris.
Campbell, J.M.
1968 A Revision of the New World Micropeplinae (Coleoptera: Staphylinidae) with a rearrangement of the world species. *Canadian Entomologist* 100(3): 225-267.
1975 A Revision of the Genera *Coprophilus* and *Cilea* (Coleoptera: Staphylinidae) of America North of Mexico. *Canadian Entomologist* 107: 175-216.
1976 A Revision of the Genera *Sepedophilus* Gistel (Coleoptera: Staphylinidae) of America North of Mexico. *Memoirs of the Entomological Society of Canada* 99. Entomological Society of Canada, Ottawa.
Campbell, J.M., M.J. Sarazin, and D.B. Lyons
1989 *Canadian Beetles (Coleoptera) injurious to crops, ornamentals, stored products, and buildings.* Research Branch, Agriculture Canada, Ottawa.
Cartwright, F.F.
1977 *A Social History of Medicine*. Longman, London.
Catts, E.P. and N.H. Haskell
1990 *Entomology and Death: a Procedural Guide.* Joyce's Print Shop, Clemson.
Census Departement
1863-1864 *Census of the Canadas, 1860-1861.* S. B. Foote, Québec.
Chandler, A.C.
1955 *Introduction to Parasitology.* General Publishing Ltd., Toronto.
Chase A.W.
1864 *Dr. Chase's recipes, or, Information for everybody: an invaluable collection of about Eight Hundred Recipes for....* A. Taylor, London, Ontario.
Chénier, R.
1991 Québec: A French Colonial Town in America, 1660-1690. Minister of Supply and Services, Ottawa.
Cherrier, G.H.
1864 *The Québec Directory for 1864-5, containing a complete list of all the inhabitants of the City, with full descriptions of their businesses and professions: and Information carefully collected in reference to Government and Public Offices, Banks, Corporate Institutions & Companies in Quebec.* John Lovell, Québec.
1870 *The Québec Directory for 1870-1, containing a complete list of all the inhabitants of the City, with full description of their businesses and professions, and Information carefully collected in reference to Government and Public Offices, Banks, Corporate Institutions & Companies in Quebec.* John Lovell, Québec.
1877 *Cherrier's directory of Québec and Levis for the year ending May 1...* M. L. A. J. Cherrier, Québec.
Cherrier, G.H. and P.M. Hamelin
1858 *The Québec Directory for 1858-9, containing a complete list of all the inhabitants of the City, with full descriptions of their businesses and professions; and information carefully collected in reference to all the banks, public offices, corporate institutions, and companies in Quebec"* P. Lamoureux Québec.
Chouinard, M.
1901 *Règlements du Conseil de Ville de Québec.* Demers & Frère, Québec, Québec.
Cloutier, C.
1996 Hygiène privée, transferts culturels et mécanisation:le cas des latrines du site Aubert de la Chesnaye à Québec. In *L'eau, l' hygiène publique et les infrastructures*, edited by Louise Pothier, pp: 17-24.
Collection mémoires vives, Groupe PGV, Montréal.
Cloutier, P.
1991 Îlot Hunt, chaîne de titres de propriété. Submitted to CELAT, Université Laval, Québec, Québec.
Conseil de ville de la cité de Québec
1856 Reglements, regles et ordures pour la regi interieure du conseil de ville de la cité de Québec. Imprimé par St. Michel et Darveau, No. 3, Rue La Montagne, Québec.
Cook, L.J.
1998 Katherine Nanny, alias Naylor: A Life in Puritan Boston. *Historical Archaeology* 32(1): 15-19.
Coope, G.R.
1970 Interpretations of Quaternary insect fossils. *Annual Review of Entomology* 15: 97-120.
1973 Tibetan Species of Dung Beetle from Late Pleistocene Deposits in England. *Nature* 245: 335-36.
1978 Constancy of insect species versus inconstancy of Quaternary environments. In *Symposia of the Royal Entomological Society of London* 9, edited by Laurence A. Mound and Nadia Waloff, pp: 176-187, Blackwell Scientific Publications, Oxford.
1996 Validations for the use of beetle remains as reliable indicators of Quaternary climates: a reply to the criticism by Johan Andersen. *Journal of Biogeography* 23: 115-121.
Coope, G.R. and J.A. Brophy
1976 Late Glacial environmental changes indicated by a coleopteran succession from North Wales. *Boreas* 1: 97-142.

Coope, G.R. and P.J. Osborne
1968 Report on the Coleopterous Fauna of the Roman Well at Barnsely Park, Gloucestershire. *Transactions of the Bristol and Gloucestershire Archaeological Society* 86: 84-87.

Copp, T.
1978 *Classe ouvrière et pauvreté, les conditions de vie des travailleurs montréalais, 1897-1929*. Boréal Express, Montréal.

Cronon, W.
1991 *Nature's Metropolis: Chicago and the Great West*. W. W. Norton & Company, New York.

Crosby, A.W.
1986 *Ecological Imperialism: The Biological Expansion of Europe, 900-1900*. Cambridge University Press, Cambridge.

Dagneau, G-H., A. Duval, A. Lebel, R. Lemoine, G. Noël, H. Provost and M. Vallieres
1983 La Ville de Québec, histoire municipale IV. *Cahiers d'histoire* 35. La société historique du Québec, Québec, Québec.

Deagan, K.
1996 Environmental and Historical Archaeology. In *Case Studies in Environmental Archaeology*, edited by Elizabeth J. Reitz, Lee A. Newsom, and Sylvia J. Scudder, pp: 359-376. Plenum Press, New York.

de la Torre-Bueno, J.R.
1978 *A Glossary of Entomology*. New York Entomological Society, New York.Department of Agriculture
1873-1878 *Census of Canada, 1870-1871*. I. B. Taylor, Ottawa.
1882-1885 *Census of Canada, 1880-1881*. R. Maclean, Ottawa.
1893-1897 *Census of Canada, 1890-1891*. King's Printers, Ottawa.

Desloges, Y.
1991 *A Tenant's Town*. Minister of Supply and Services Canada, Ottawa.

deVarennes, P.
1994 *Le Site de l'Îlot Hunt: rapport d'une surveillance archéologique sur le site de l'Îlot Hunt*. Submitted to CELAT, Université Laval, Québec and Ville de Québec, Québec.

Dirrigl Jr., F.J. and B. Greenberg
1995 The Utility of Insect Remains to Assessing Human Burials: A Connecticut Case Study. *Archaeology of Eastern North America* 23: 1-7.

Downie, N.M. and R.H. Arnett, Jr.
1996 *The beetles of northeastern North America*. Volume II. Gainesville, The Sandhill Crane Press.

Driscoll, L.H.
1994 *Privy Unto What Has Passed: The Application and Limitations of Parasitology for Eastern North American Historical Sites*. Unpublished M.A. Thesis, University of Massachusetts, Boston.
1995 *Parasitological Analysis: 1995 Samples from Central Artery/ Third Harbor Tunnel Project, Boston, Massachusetts, Cross Street- Feature 4*. Submitted to Timelines, Inc., Littleton, Massachusetts.

Drolet, A.
1965 La Ville de Québec, histoire municipale II. *Cahiers d'histoire* 17. La société historique du Québec, Québec, Québec.
1967 La Ville de Québec, histoire municipale III. *Cahiers d'histoire* 19. La société historique du Québec, Québec, Québec.

Dubé, J.
1995 *Le site de l'Îlot Hunt à Québec (CeEt-110), rapport annuel d'activities, quatrième campagne de fouilles archéologiques (1994)*. Submitted to CELAT, Université Laval, Québec and Ville de Québec, Québec.

Duffin, J.
1993 *Langstaff: A Nineteenth-Century Medical Life*. University of Toronto Press, Toronto.

Duffy, J.
1990 *The Sanitarians: A History of Public Health*. University of Illinois Press, Urbana.

Elias, S.A.
1985 Paleoenvironmental interpretation of Holocene insect fossil assemblages from four high-altitude sites in the Front Range,
Colorado, U.S.A. *Arctic and Alpine Research* 17: 31-48.
1994 *Quaternary Insects and Their Environments*. Smithsonian Institution Press, Washington.

Ervynck, A.K., K. Desender, M. Pieters, and J. Bungeers.
1994 Carabid beetles as palaeo-ecological indicators in archaeology. In *Carabid Beetles: Ecology and Evolution*, edited by Konjev Desender et al., pp: 261-266. Kluwer Academic Publishers, Netherlands.

Evans, J.G.
1978 *Introduction to Environmental Archaeology*. Cornell University Press, Ithaca.

Faulkner C.T.
1991 Prehistoric Diet and Parasitic Infection in Tennessee: Evidence from the Analysis of Desiccated Human Paleofeces. *American Antiquity* 56(4) 687-700.

Follwell, J.H.
1952 Notes on some Ptinidae of British Columbia (Coleoptera). *Entomological Society of British Columbia, Proceedings* 48: 60-63.

Fortin, C.
1996 Les macrorestes végétaux de l'ilot Hunt (fouilles de 1993, site CeEt 100, Québec). In *Des vestiges d'une arrière-cour à l'histoire de l'hygiène publique à Québec au XIXe siècle : la troisième campagne de fouilles archéologiques à l'îlot Hunt (1993)*. Unpublished M.A. thesis by Manon Goyette, Université Laval, Québec, 1999, Annexe C.

Frank, J.H.
1975 A revision of the New World species of the genus *Erichsonius* Fauvel (Coleoptera: Staphylinidae). *The Coleopterists Bulletin* 29(3): 177-203.
1982 A revision of the New World species of the genus *Neobisnius* Ganglbauer (COLEOPTERA: STAPHYLINIDAE: STAPHYLININAE). *Occasional Papers of the Florida State Collection of Arthropods* 1. Florida Department of Agriculture and Consumer Services, Gainesville.

Fries, C.R., D.K. Beidleman, and J.F. Custer
1990 Analysis of parasites from 19th century contexts, Wilmington, Delaware. *North American Archaeologist* 11(1): 17-27.
Garrow, P.
1996 Archaeological Investigations of the Courthouse Block, Knoxville, Tennessee. Paper presented at the 1996 Annual Meeting of the Society for Historical Archaeology, Cinicinnati, Ohio.
Geismar, J.H.
1993 Where is nightsoil? Thoughts on an urban privy. *Historical Archaeology* 27(2): 57 -70.
Gibson, A.
1924 The occurrence of the ptinid beetle, *Niptus hololeucus* Fald., in North America. *Canadian Entomologist* 56: 74-76.
Gibson, L.P.
1969 Monograph of the Genus *Curculio* in the New World (Coleoptera: Curculionidae) Part I. United States and Canada. *Miscellaneous Publications of the Entomological Society of America* 6: 241-285.
Gill, G.W. and D.W. Owsley
1985 Electron microscopy of Parasite Remains on the Pitchfork Mummy and Possible Social Implications. *Plains Anthropologist* 30: 45-50.
Girling, M.
1978 The Application of Fossil Insect Studies to the Somerset Levels In *Research Problems in Zooarchaeology*, edited by Don R. Brothwell, K. D. Thomas and Juliet Clutton-Brock, pp: 85-90. Occasional Paper No. 3, Institute of Archaeology, London.
1984 Eighteenth Century Records of Human Lice (Pthiraptera, Anoplura) and Fleas (Siphonaptera, Pulcidae) in the City of London. *Entomologist's Monthly Magazine* 120: 207-210.
Gordon, R.D.
1983 Studies on the genus *Aphodius* of the United States and Canada (Coleoptera: Scarabaeidae). VII. Food and Habitat; Distribution; Key to Eastern Species. *Proceedings of the Entomological Society of Washington* 85(4): 633-652.
Goulet, D.
1996 Des bureaux d'hygiène municipaux aux unités sanitaires. Le Conseil d'hygiène publique de la province de Québec et la structuration d'un système de santé publique 1886-1926. *Revue d'histoire d'Amérique française*, 49(4): 409-520.
Goulet, D. and O. Keel
1993 Les hommes-relais de la bactériologie en territoire québécois et l'introduction de nouvelles practiques diagnostiques et thérapeutiques, (1890-1920). *Revue d'histoire d'Amérique française* 46(3): 417-442.
Goulet, D. and Paradis A.
1992 *Trois siècles d'histoire médicale au Québec: chronologie des institutions et des pratiques.* VLB éditeur, Montréal.
Goyette, M.
1994 *Le site de l'îlot Hunt a Québec, rapport annuel d'activités, troisième campagne de fouilles archéologiques (1993).* Submitted to Université Laval, Québec and Ville de Québec, Québec.
1999 Des vestiges d'une arrière-cour à l'histoire de l'hygiène publique à Québec au XIXe siècle : la troisième campagne de fouilles archéologiques à l'îlot Hunt (1993). Unpublished M.A. thesis, Département d'histoire, Université Laval, Québec.
Greig, J.
1981 The Investigation of a Medieval Barrel Latrine. *Journal of Archaeological Science* 8: 265-282.
1982 Garderobes, Sewers, Cesspits and Latrines. *Current Archaeology* 85: 49-52
Groupe Harcart, Le
1989 *Inventaire archéologique, îlots Bell et Hunt.* Submitted to Ville de Québec, service de l'urbanisme, Québec.
Guérard, F.
1996 L'hygiène publique au Québec de 1887 à 1939 : centralisation, normalisation et médicalisation. *Recherches sociographiques*, 37(2): 203-227.
Hakbijl, T
1986 Remains of Insects. In *Amsterdam Project, Annual Report of the VOC-Ship "Amsterdam" Foundation 1985*, edited by Jerzy H.G. Gawronski, pp: 72-73. Amsterdam Project, Amsterdam.
1986 Insect remains: unadulterated Cantharidum and tobacco from the West Indies. In *Amsterdam Project, Annual Report of the VOC-Ship "Amsterdam" Foundation 1986*, edited by Jerzy H.G. Gawronski, pp: 92-94. Amsterdam Project, Amsterdam.
Hakbijl, T. and M. de Groot
1997 Insect Remains from Willem Barents' 1596 Arctic Exploration Preserved in "Het Behouden Huys", Novaya Zemlya- with notes on the medicinal use of Cantharides. *Quaternary Proceedings* 5: 129-134.
Hall, A.R., A.K.G. Jones, and H.K. Kenward
1983 Cereal Bran and Human Faecal Remains from Archaeological Deposits - Some Preliminary Results. In *Site, Environment and Economy*, edited by Bruce Proudfoot, pp: 85-104. BAR International Series 173. British Archaeological Reports, Oxford.
Hall, A.R. and H.K. Kenward
1976 Biological evidence from the usage of Roman riverside warehouses at York. *Britannia* 7: 274-276.
1990 Environmental Evidence from the Colonia: General Accident and Rougier Street. *The Archaeology of York, The Past Environment of York* 14/6. Council for British Archaeology for York Archaeological Trust, London.
Hall, A.R., H.K. Kenward, and D. Williams
1980 Environmental Evidence from Roman Deposits in Skeldergate. *The Archaeology of York, The Past Environment of York* 14/3. Council for British Archaeology for York Archaeological Trust, London.
Hall, A.R., H.K. Kenward, D. Williams, and J.R.A. Greig
1983 Environment and Living Conditions at Two Anglo-Scandinavian Sites. *The Archaeology of York, The Past Environment of York* 14/4. Council for British Archaeology for York Archaeological Trust, London.
Hall, H.
1977 A Paleoscatological Study of Diet and Disease at Dirty Shame Rockshelter, Southeast Oregon. *Tebiwa* 8: 1-15.

Halstead, P.
1990 Quantifying Sumerian Agriculture- Some Seeds of Doubt and Hope. *Bulletin on Sumerian Agriculture* 5: 187-195.
Hammond, P.M.
1976 A review of the genus *Anotylus* C. G. Thomson (Coleoptera : Staphylinidae). *Bulletin of the British Museum (Natural History) Entomology* 33(2): 139-187.
Hansen, M.
1996 Katalog over Danmarks biller, Catalogue of the Coleoptera of Denmark. *Entomologiske Meddelelser* 64: 1-231. Apollo Books, Copenhagen.
Hare, J., M. Lafrance, and D-T. Ruddel
1987 *Histoire de la ville de Québec 1608-1871.* Boreal, Montréal.
Hatch, M.H.
1957 Beetles of the Pacific Northwest II: Staphyliniformia. *University of Washington Publications in Biology* 16, University of Washington Press, Seattle.
Hinton, H.E.
1945a *A monograph of the Beetles associated with stored products.* Jarrold and Sons, Ltd., London.
1945b Histeridae associated with stored products. *Bulletin of Entomological Research* 35: 309-340.
Hodder, I.
1986 *Reading the Past.* Cambridge University Press, Cambridge.
1987 The contextual analysis of symbolic meanings. In *The archaeology of contextual meanings - New directions in archaeology*, edited by Ian Hodder, pp: 1-10. Cambridge University Press, Cambridge.
Horion, A.
1949 *Faunistik Der Mitteleuropaïschen Käfer* 2. Vittorio Klostermann, Frankfurt.
1963 *Faunistik der mitteleuropäischen Käfer* 9. Überlingen-Bodensee.
Horne, P.D.
1985 A Review of the Evidence of Human Endoparasitism in the pre-Columbian New World Through the Study of Coprolites. *Journal of Archaeological Science* 12: 299-310.
Horne, P.D. and J.A. Tuck.
1996 Archaeoparasitology at a 17th Century Colonial Site in Newfoundland. *Journal of Parasitology* 82(3): 512-515.
Howson, J.
1993 The Archaeology of 19th-Century Health and Hygiene at the Sullivan Street Site, New York City. *Northeast Historical Archaeology* 22: 137-160.
Huckett, H.C. and J.R. Vockeroth
1987 Muscidae. In *Manual of Nearctic Diptera, Volume 2*, edited by J. Frank McAlpine, Bobby V. Peterson, Guy E. Shewell, Herb J. Teskey, J. Richard Vockeroth, and D. Monty Wood, pp: 1115-1131, Research Branch, Agriculture Canada, Monograph 28, Minister of Supply and Services, Ottawa.
Jones, A.K.
1982 Human parasite remains: prospects for a quantitative approach. In *Environmental Archaeology in the Urban Context*, edited by Allan R. Hall and Harry K. Kenward, pp: 66-70. Research Report 43, Council for British Archaeology, London.
1985 Trichurid Ova in Archaeological Deposits: Their value as Indicators of Ancient Faeces. In *Palaeobiological Investigations: Research Design, Methods and Data Analysis*, edited by Nick R.J. Fieller, David D. Gilbertson, and Nick G.A. Ralph, pp: 105-119. BAR International Series 266. British Archaeological Reports, Oxford.
Jones, G., K. Wardle, P. Halstead, and D. Wardle
1986 Crop Storage at Assiros. *Scientific American* 254: 96-103.
Jones, M.
1986 *England before Domesday*. Batsford, London.
Jones, O., C. Sullivan and G.L. Miller
1985 *The Parks Canada glass glossary for the description of containers, tableware, flat glass and closures.* Parks Canada, Ottawa.
Kellogg, V.L. and G.F. Ferris
1915 The Anoplura and Mallophaga of North American Mammals. *Leland Stanford Junior University Publications.* Stanford University, California.
Kenward, H.K.
1974 Methods for palaeo-entomology on site and in the laboratory. *Science and Archaeology* 13: 16-24.
1975a Pitfalls in the Environmental Interpretation of Insect Death Assemblages. *Journal of Archaeological Science* 2: 85-94.
1975b The Biological and Archaeological Implications of the Beetle *Aglenus brunneus* (Gyllenhal) in Ancient Faunas. *Journal of Archaeological Science* 2: 63-69.
1976a Further Archaeological records of *Aglenus brunneus* (Gyll.) in Britain and Ireland, Including Confirmation of Its Presence in the Roman Period. *Journal of Archaeological Science* 3: 275-77.
1976b Reconstructing ancient ecological conditions from insect remains: some problems and an experimental approach. *Ecological Entomology* 1: 7-17.
1978a The Analysis of Archaeological Insect Assemblages: A New Approach. *The Archaeology* of York, Principles and Methods 19/1. Council for British Archaeology for York Archaeological Trust, London.
1978b The value of insect remains as evidence of ecological conditions on archaeological sites. In *Research Problems in Zooarchaeology*, edited by Don R. Brothwell, K.D. Thomas and Juliet Clutton-Brock, pp: 25-38. Occasional Paper No. 3, Institute of Archaeology, London.
1982 Insect communities and death assemblages, past and present. In *Environmental archaeology in the urban context*, edited by Allan R. Hall and Harry K. Kenward, pp: 71-78. Council for British Archaeology Research Report 43, London.
1985 Outdoors – Indoors? The Outdoor Component of Archaeological Insect Assemblages. In *Palaeobiological Investigations – Research Design, Methods and Data Analysis*, edited by Nick R.J. Fieller, David D. Gilbertson and Nick G.A. Ralph, pp: 97-106. BAR International Series 266, Oxford, England.
1997 Synanthropic decomposer insects and the size, remoteness and longevity of archaeological occupation sites: applying concepts from biogeography to past 'islands' of human occupation. *Quaternary Proceedings* 5: 135-151.

1999 Public lice (*Pthirus pubis* L.) were present in Roman and Medieval Britain. *Antiquity* 73: 911-915.

Kenward, H. and E. Allison

1994 Rural origins of the urban insect fauna. In *Urban-Rural Connexions: Perspectives from Environmental Archaeology*, edited by Allan R. Hall and Harry K. Kenward , pp: 55-77. Oxbow Monograph 47, Association for Environmental Archaeology, Exeter.

Kenward, H. and A. Hall

1997 Enhancing Bioarchaeological Interpretation Using Indicator Groups: Stable Manure as a Paradigm. *Journal of Archaeological Science* 24: 663-673.

Kenward, H.K., A.R. Hall, and A.K.G. Jones

1980 A tested set of techniques for the extraction of plant and animal macrofossils from waterlogged archaeological deposits. *Science and Archaeology* 22: 3-14.

Kenward, H. and F. Large

1998 Insects in Urban Waste Pits in Viking York: Another Kind of Seasonality. *Environmental Archaeology* 3: 35-53.

Kenward, H.K. and D. Williams

1979 Biological Evidence from the Roman Warehouses in Coney Street. *The Archaeology of York, The Past Environment of York* 14/2. Council for British Archaeology for York Archaeological Trust, London.

Kislev, M.E.

1991 Archaeobotany and storage archaeoentomology. In *New Light on Early Farming: Recent Developments in Palaeoethnobotany*, edited by Jane M. Renfrew, pp: 121-136. Edinburgh University Press, Edinburgh.

Klimaszewski, J.

1984 A revision of the genus *Aleochara* Gravenhorst of America North of Mexico (Coleoptera: Staphylinidae: Aleocharinae). *Memoirs of the Entomological Society of Canada* 129. Entomological Society of Canada, Ottawa.

Lachapelle, S.

1891 Manual of Hygiene for the Use of School and Families Based on the Instructions of the Board of Health of the Province of Québec. Desaulniers & Leblanc, Montréal.

Lafleur C.

1999 La céramique institutionnelle du dépôt de la firme A.E. Vallerand. Unpublished M.A. thesis, Département d'histoire, Université Laval, Québec.

Laframboise, Y.

1991 La fonction residentielle de Place-Royale, 1820-1860 Synthèse. *Collection patrimoines* 70. Les publications du Québec, Québec.

Landin, B-O.

1961 Ecological Studies on Dung Beetles (Coleoptera: Scarabaeidae). *Opuscula Entomologica Supplementum* 19: 1-228. Entomologiska sallskapet i Lund, Lund.

L'Anglais, P-G.

1993 *Le site de l'îlot Hunt, rapport de la deuxième campagne de fouilles (1992).* Submitted to CELAT, Université Laval, Québec and Ville de Québec, Québec.

1996 L'hygiène urbaine à Québec au temps de la Nouvelle France. In *L'eau, l' hygiène publique et les infrastructures*, edited by Louise Pothier, pp: 9-16. Collection mémoires vives, Groupe PGV, Montréal.

1998 Le site de l'îlot Hunt, rapport de la deuxième campagne de fouilles (1992). *Cahiers d'archéologies du CELAT* 2. Université Laval, Québec.

Larocque, P.

1970 La condition socio-économiques des travaillers de la ville de Québec - 1894-1914. Unpublished M. A. thesis, Département d'histoire, Université Laval, Québec.

Larsen, E.L.

1994 A Boardinghouse Madonna-Beyond the Aesthetics of a Portrait Created Through Medicine Bottles. *Historical Archaeology* 28(4): 68-79.

Lavoie, C., Elias, S.A. and Payette, S.

1997 Holocene fossil beetles from a treeline peatland in subarctic Quebec. *Canadian Journal of Zoology* 75: 227-236.

Leavitt, J.W.

1982 *The Healthiest City: Milwaukee and the politics of Health Reform.* Princeton University Press, Princeton, New Jersey.

Leclerc M.

1992 *Le site de l'îlot Hunt, première campagne de fouilles (1991), Rapport annuel d'activities.* Submitted to CELAT, Université Laval, Québec, and Ville de Québec, Québec.

1995 *L'appropriation de l'espace et urbanisation d'un site de la basse ville de Québec. La première campagne de fouilles à l'îlot Hunt (1991).* Unpublished M. A. thesis, Département d'histoire, Université Laval, Québec.

1998 L'appropriation de l'espace et urbanisation d'un site de la basse ville de Québec. Rapport de la première campagne de fouilles à l'îlot Hunt (1991), *Cahiers d'archéologie du CELAT* 1. Université Laval, Québec.

Lindroth, C.H.

1957 *The faunal connections between Europe and North America.* John Wiley and Sons, New York.

1963 The ground beetles (Carabidae excl. Cincindelinae) of Canada and Alaska. *Opuscula Entomologica Supplementum* 24: 201-408. Entomologiska sallskapet i Lund, Lund.

1966 The ground beetles (Carabidae excl. Cincindelinae) of Canada and Alaska. *Opuscula Entomologica Supplementum* 29: 409-648. Entomologiska sallskapet i Lund, Lund.

1969 The ground beetles (Carabidae excl. Cincindelinae) of Canada and Alaska. *Opuscula Entomologica Supplementum* 34: 945-1192. Entomologiska sallskapet i Lund, Lund.

Lorentzen, B. and A.M. Rørdam

1989 Investigation of Faeces from a Mummified Eskimo Woman. *Meddelelser om Grømland, Man and Society* 12: 139-143.

Lowe, J.J. and M.J.C. Walker
1984 *Reconstructing Quaternary Environments*. Longman Scientific and Technical, Essex.

MacDougall, H.
1982 The Genesis of Public Health Reform in Toronto, 1869-1890. *Revue d'histoire urbain* 10(3): 1-9.

Mackay, R.W.S.
1848-1849 *Mackay's Québec directory*. R. W. S. Mackay, Québec.

Marcotte
1900 *Annuaire Marcotte de Québec métopolitain*. Polk, Québec.

Marshall, S.A. and O.W. Richards
1987 Sphaeroceridae. In *Manual of Nearctic Diptera, Volume 2*, edited by J. Frank McAlpine, Bobby V. Peterson, Guy E. Shewell, Herb J. Teskey, J. Richard Vockeroth, and D. Monty Wood, pp: 993-106, Research Branch, Agriculture Canada, Monograph 28, Minister of Supply and Services, Ottawa.

Matthews, J.V.
1981 Tertiary land bridges and their climate: Backdrop for the development of the present Canadian insect fauna. *Canadian Entomologist* 112: 1089-1103.
1983 A Method of Comparison of Northern Fossil Insect Assemblages. Géographie physique et Quaternaire 37(3): 297-306.

McAlpine, J.F., B.V. Peterson, G.E.Shewell, H.J. Teskey, J.R.Vockeroth, and D.M. Wood
1981 Introduction. In *Manual of Nearctic Diptera, Volume 1*, edited by J. Frank McAlpine, Bobby V. Peterson, Guy E. Shewell, Herb J. Teskey, J. Richard Vockeroth, and D. Monty Wood, pp: 1-7. Monograph 27, Minister of Supply and Services, Ottawa.

Mead, L.
1997 *Management Summary, Archeological Evaluation, and Data Recovery at Boston African American National Historic Site, Boston, Massachusetts*. National Park Service, Submitted to Boston national Historical Park.

Moore, J.G., A.W Grundmann, H.J. Hall, and G.F. Fry
1974 Human fluke infection at Glen Canyon AD 1250. *American Journal of Physical Anthropology* 41: 115-118.

Moore, R.
1874 *The Artizans' Guide and Everybody's Assistant Containing Over Three Thousand New and Valuable Receipts and Tables in Almost Every Branch of Business Connected with Civilized Life, from the Household to the Manufactory*. John Lovell, Montréal.

Morgan, A.V.
1987 Late Wisconsin and early Holocene paleoenvironments of east-central North America based on assemblages of fossil Coleoptera. In *North America and adjacent oceans during the last deglaciation*, edited by William F. Ruddiman and H. E. Wright Jr., pp: 353-370. *The Geology of North America* K-3. Geological Society of America, Boulder, Colorado.

Morgan, A.V.and A. Morgan.
1980a Beetle Bits- The Science of Paleoentomology. *Geoscience Canada* 7(1): 22-29.
1980b Faunal Assemblages and Distributional Shifts of Coleoptera During the Late Pleistocene in Canada and in the Northern United States. *Canadian Entomologist* 112: 1105-1128.
1987 Paleoentomology- Towards the Next Decade. *Episodes* 10(1): 38-40.

Moussette, M. and R. Auger
1997 Pour une archéologie contextuelle de la ville. Creuser la mémoire de Québec. In *Les espaces de l'identité*, edited by Laurier Turgeon, Jocelyn Létourneau and Khadyatoulah Fall, pp: 225-236. Presses de l'Université Laval, Québec.

Mrozowski, S.A.
1987 Exploring New England's Evolving Urban Landscape. In *Living in Cities: Current Research in Archaeology*, edited by Edward Staski, pp: 1-9. Society for Historical Archaeology, Pleasant Hill, California.
1988 Historical Archaeology as Anthropology. *Historical Archaeology* 22(1): 18-24.

Mrozowski, S.A., E.L. Bell, Mary C. Beaudry, D.B. Landon, and G.K. Kelso
1989 Living on the Boott: health and well being in a boardinghouse population. *World Archaeology* 21(2): 298-319.

Mulsant, M.E.
1857 *Histoire Naturelle des coleoptères de France*. Magnin, Blanchard et Cie., Lyons.

O'Brien, C.W.
1981 The Larger (4.5+ mm) *Listronotus* of America, North of Mexico (Cylindrorhininae, Curculionidae, Coleoptera). *Transactions of the American Entomological Society* 107: 69-123.

O'Connor, T.P.
1998 Environmental Archaeology: a Matter of Definition. *Environmental Archaeology* 2: 1-6.

Osborne, P.J.
1969 An insect fauna of Late Bronze Age date from Wilsford, Wiltshire. *Journal of Animal Ecology* 38: 555-566.
1971 An Insect Fauna from the Roman Site at Alcester, Warwickshire. *Britannia* 2: 156-165.
1981 The Investigation of a Medieval Barrel Latrine. *Journal of Archaeological Science* 8: 265-282.
1983 An Insect Fauna from a Modern Cesspit and its Comparison with Probable Cesspit Assemblages from Archaeological Sites. *Journal of Archaeological Science* 10: 453-463.
1988 A Late Bronze Age Fauna from the River Avon, Warwickshire, England: Its Implications for the Terrestrial and Fluvial Environment and for Climate. *Journal of Archaeological Science* 15: 715-727.

Painchaud, A.
1993 Paléogéographie de la pointe de Québec (Place-Royale). *Collection patrimoines* 83. Les publications du Québec, Québec.

Paquet, N.
1990 Les élites de Paris et le problème urbain des fosses d'aisances et des voiries dans la première moitié du XIXe siècle. Unpublished M. A. thesis, Département d'histoire, Université Laval, Québec.

Pals, J.P. and T. Hakbijl
1992 Weed and insect infestation of a grain cargo in a ship at the Roman fort of Laurium in Woerden

(Province of Zuid-Holland). *Review of Palaeobotany and Palynology* 73: 287-300.

Panagiotakopulu, E. and P.C. Buckland

1991 Insect Pests of Stored Products from Late Bronze Age Santorini, Greece. *Journal of Stored Product Research* 27(3): 179-184.

1999 *Cimex lectularius* L., the common bed bug from Pharonic Egypt. *Antiquity* 73: 908-911.

Parry, R.H.

1986 The Systematics and Biology of the Flea Beetle Genus *Crepidodera* Chevrolat (Coleoptera:Chrysomelidae) in America North of Mexico. *Insecta Mundi* 1(3): 159-196.

Peterson, R.T. and M. McKenny

1968 *A field guide to wildflowers, Northeastern and North-central North America.* Houghton Mifflin Company, New York.

Pierre-Deschênes, C.

1981 Santé publique et organisation de la profession médicale au Québec. *Revue d'histoire d'Amérique française*, 35(3): 355-375.

Pothier, L. (editor)

1996 *L'eau, l' hygiène publique et les infrastructures.* Collection mémoires vives, Groupe PGV, Montréal.

1996 Réseaux d'eau potable et d'eaux usées : L'hygiène publique dans la société montréalaise (1642-1910) In *L'eau, l' hygiène publique et les infrastructures*, edited by Louise Pothier. Collection mémoires vives, Groupe PGV, Montréal.

Quebec Official Gazette

1885 Resolutions adopted by the Central Board of Health on the 10th, 11th, and 12 th of September 1885, Province of Quebec. Quebec Official Gazette, Quebec.

1891 By-Laws and Regulations of the Board of Health and the Province of Quebec made under the Authority of Act 54, Vict., Chap. 27, 13th June. Quebec Official Gazette, Quebec.

Rapport de L'Inspecteur des Chemins

1852 Rapport de L'Inspecteur des Chemins sur la Canalisation de la Cité de Québec. Bureau et Marcotte, Québec.

Rapport du Conseil d'hygiène de la Province de Québec

1895 Rapport du Conseil d'hygiène de la Province de Québec. Charles François Langlois, Montréal.

Ratcliffe, B.C.

1972 The Natural History of *Necrodes surinamensis* (Fabricius) (Coleoptera: Silphidae). *Transactions of the American Entomological Society* 98(4): 309-410.

1991 The Scarab Beetles of Nebraska. *Bulletin of the University of Nebraska State Museum*, Volume 12. University of Nebraska, Lincoln.

Recherches Arkhis Inc., Les

1987 Annexe 2. Les modes de vie de la population de Place-Royale entre 1820-1859: l'information archéologique. *Collection patrimoines* 66, Les Publications du Québec, Québec.

1990 Annexe 1. Les modes de vie de la population de Place-Royale entre 1820-
1859. Annexe 1. *Collection patrimoines* 69, Les publications du Québec, Québec.

Reinhard, K.J.

1988 Cultural Ecology of Prehistoric Parasitism on the Colorado Plateau as Evidenced by Coprology. *American Journal of Physical Anthropology* 77: 355-366.

1989 Parasitological Analysis of Latrine Soils from the Boott Mills Boardinghouse Site, Lowell, Massachusetts In *Interdisciplinary Investigations of the Boott Mills, Lowell, Massachusetts, Volume III, The Boarding House System as a Way of Life*, edited by Mary C. Beaudry, and Stephen A. Mrozowski, pp: 271-272. Cultural Resources Management Studies No. 21.U.S. Department of the Interior, National Park Service, North Atlantic Regional Office, Boston, Massachusetts.

1990 Archaeoparasitology in North America. *American Journal of Physical Anthropology* 82:145-163.

1992 Parasitology as an Interpretive Tool in Archaeology. *American Antiquity* 57(2): 231-245.

1994 Sanitation and Parasitism at Harpers Ferry, West Virginia. *Historical Archaeology* 28(4): 62-67.

Reinhard, K.J., R.H. Hevly, and G.A. Anderson

1987 Helminth remains from prehistoric Indian coprolites on the Colorado plateau. *Journal of Parasitology* 73(3): 630-639.

Reinhard, K.J., S.A. Mrozowski and K.A. Orloski

1986 Privies, Pollen and Seeds: A Biological Nexus in Historic Archeology. *MASCA Journal* 4 (1): 31-36.

Report of a Meeting of the Montreal Natural History Society

1885 *Important Health Matters. Dr. Baker Edwards on the disposal of night-soil. Cremation Recommended and Charcoal Closets for Dwellings.* Reprinted from the 3rd February 1885 *Montreal Herald.*

Roberts, N.

1989 *The Holocene: an Environmental History.* Blackwell, Cambridge.

Robinson, W.H.

1996 *Urban Entomology: Insect and mite pests in the human environment.* Chapman and Hall, London.

Ruddel, D.T.

1981 *Québec City 1765-1831: The Evolution of a Colonial Town.* Unpublished Doctoral thesis, Département d'histoire, Université Laval, Québec.

1991 *Québec, 1765-1832.* Musée canadien des civilisations, Hull.

Sadler, J.

1990 Records of Ectoparasites on Human's and Sheep from Viking Age Deposits on the Former Western Settlement of Greenland. *Journal of Medical Entomology* 27(4): 628-631.

1991 Beetles, Boats, and Biogeography: Insect Invaders of the North Atlantic. *Acta Archaeologica* 61: 199-211.

Sadler, J.P. and P. Skidmore

1995 Introductions, Extinctions or Continuity? Faunal Change in the North Atlantic Islands. In *Ecological Relations in Historical Times* edited by Robin Butlin, and Neil Roberts, pp: 206-223. Blackwell, Oxford.

Sailer, R.I.

1983 History of Insect Introductions. In *Exotic Plant Pests and North American Agriculture,* edited by Charles L. Wilson and Charles L. Graham, pp: 15-37. Academic Press, New York.

Saucier, R.

1958 L'Hygiène à Québec sous le régime français. Thèse de doctorat, Université Laval, Québec.

Schelvis, J.
1987 Some aspects of research on mites (Acari) in archaeological samples. *Palaeohistoria* 29: 211-218.
1994 Caught between the Teeth. A Review of Dutch Finds of Archaeological Remains of Ectoparasites in Combs. *Proceedings of the section Experimental and Applied Entomology of the Netherlands Entomological Society (N.E.V.)* 5: 131-132.
1997 Mites in the background: Use and origin of remains of mites (Acari) in Quaternary deposits. *Quaternary Proceedings* 5: 233-236.

Schwert, D.P.
1996 Effect of Euro-American Settlement on an Insect Fauna: Paleontological Analysis of the Recent Chitin Record of Beetles (Coleoptera) from Northeastern Iowa. *Annals of the Entomological Society of America* 89(1): 53-63.

Schwert, D.P., and A.C. Ashworth
1988 Late Quaternary history of the northern beetle fauna of North America: A synthesis of the evidence. *Memoirs of the entomological Society of Canada* 144: 93-107.

Schwert, D.P., T.W. Anderson, A. Morgan, A.V.Morgan, and P. F. Karrow
1985 Changes in Late Quaternary Vegetation and Insect Communities in Southwestern Ontario. *Quaternary Research* 23: 205-226.

Seevers, C.H.
1978 A generic and Tribal Revision of the North American Aleocharinae (Coleoptera: Staphylinidae). *Fieldiana Zoology* 71: 1-289.

Selander, R.B.
1960 Bionomics, Systematics, and Phylogeny of *Lytta*, a Genus of Blister Beetles (Coleoptera, Meloidae). *Illinois Biological Monographs* 28, University of Illinois Press, Urbana.

Shryock, R.H.
1966 *Medicine in America: Historical Essays*. Johns Hopkins Press, Baltimore.

Simoneau, D.
1987 Évaluation préliminaire du potential archéologique du site de l'îlot Hunt, Québec. Submitted to Service de l'urbanisme, Ville de Québec, Québec.

Simoneau, D. and S. Rouleau
1988 Rapport de surveillance archéologique : rues Sault-au-Matelot, Saint-Jacques et Saint-Antoine. Submitted to Division de Vieux-Québec et du Patrimoines, Québec, Québec.

Slater, J.A. and R.M. Baranowski
1978 *How to know the true bugs (Hemiptera-Heteroptera)*. Wm. C. Brown Publishers, Dubuque, Iowa.

Smetana, A.
1965 Staphylini und Quediini (Col., Staphylinidae) von Newfoundland Südost-Labrador und Nova Scotia. *Acta Entomologica Fennica* 20: 1-60. Suomen Hyonteistieteellinen Seura, Helsinki.
1971 Revision of the tribe Quediini of America North of Mexico (Coleoptera: Staphylinidae). *Memoirs of the Entomological Society of Canada*, 79. Entomological Society of Canada, Ottawa.
1982 Revision of the Subfamily Xantholininae of America North of Mexico (Coleoptera: Staphylinidae). *Memoirs of the Entomological Society of Canada* 120. Entomological Society of Canada, Ottawa.
1988 Review of the Family Hydrophlidae of Canada and Alaska (Coleoptera). *Memoirs of the Entomological Society of Canada* 142. Entomological Society of Canada, Ottawa.
1995 Rove beetles of the subtribe Philonthina of America North of Mexico (Coleoptera: Staphylinidae) classification, phylogeny and taxonomic revision. *Memoirs on Entomology* 3. Associated Publishers, Gainesville.

Stone, A., C.W. Sabrosky, W.W. Wirth, R.H. Foote, and J.R. Coulson
1965 *A Catalogue of the Diptera of America North of Mexico*. Agricultural Research Service, United States Department of Agriculture, Washington, D.C.

Sveinbjarnardóttir, G. and P.C. Buckland
1983 An uninvited guest. *Antiquity* 57: 127-130.

Tétreault, M.
1983 Les maladies de la misère, aspects de la santé publique à Montréal 1880-1914. *Revue d'histoire de l'amérique française* 36(4): 507-526.

Tomes, N.
1990 The Private Side of Public Health: Sanitary Science, Domestic Hygiene, and The Germ Theory, 1870-1920. *The Bulletin of the History of Medicine* 64: 509-539.

Trépanier, M.
L'eau, la technique et l'urbain : l'ingénieur n'est jamais seul dans l'uinvers des infrastructures urbaines. In *L'eau, l' hygiène publique et les infrastructures*, edited by Louise Pothier. Collection mémoires vives, Groupe PGV, Montréal.

Vaurie, P.
1955 A revision of the genus *Trox* in North America (Coleoptera, Scarabaeidae). *Bulletin of the American Museum of Natural History* 106: 1. American Museum of Natural History, New York.

Veit, R.
1996 "A Ray of Sunshine in the Sickroom": Archaeological Insights in Late 19th- and Early 20th-Century Medicine and Anesthesia. *Northeast Historical Archaeology* 25: 33-50.

Vickery, V.R. and D.K. McE. Kevan
1985 The Grasshoppers, Crickets, and Related Insects of Canada and Adjacent Regions. Ulonata: Dermaptera, Cheleutoptera, Notoptera, Dictuoptera, Grylloptera, and Orthoptera. *The Insects and Arachnids of Canada* 14. Biosystematics Research Institute, Ottawa.

Warnock, P.J. and K.J. Reinhard
1992 Methods for Extracting Pollen and Parasite Eggs from Latrine Soils. *Journal of Archaeological Science* 19: 261-264.

Werner, F.G.
1964 A revision of the North American species of *Anthicus*, s. str. (Coleoptera: Anthicidae). *Miscellaneous Publications of the Entomological Society of America* 4: 195-242.

Whitney, G.G.
1994 *From Coastal Wilderness to Fruited Plain: A History of Environmental Change in Temperate North America, 1500 to the Present.* Cambridge University Press, Cambridge.

Wirth, W.W., W.N. Mathis, and J.R.Vockeroth
1987 Ephydridae. In *Manual of Nearctic Diptera, Volume 2*, edited by J. Frank McAlpine, Bobby V. Peterson, Guy E. Shewell, Herb J. Teskey, J. Richard Vockeroth, and D. Monty Wood, pp: 1027-1047, Research Branch, Agriculture Canada, Monograph 28, Minister of Supply and Services, Ottawa.

Wood, S.L.
1982 Checklist of Bark Beetles (Coleoptera: Scolytidae). *Great Basin Naturalist Memoirs* 6. The University, Provo.

Yamin, R.
1997 New York's Mythic Slum: Digging lower Manhattan's infamous Five Points. *Archaeology* 50(2): 44-53.
1998 Lurid Tales and Homely Stories of New York's Notorious Five Points. *Historical Archaeology* 32(1): 74-85.

Acknowledgements

The data presented here are the results of my doctoral research at Université Laval, Québec, Canada, and many people assisted me in this endeavour. I would first like to thank my thesis advisor Réginald Auger and co-advisor Marcel Moussette for their guidance, advice and support. Marie-Claude Boileau, Pierre Bouchard, Paul Buckland, Lise Jodoin, and Aleš Smetana also aided me with many, many tasks throughout my doctoral program, and this was greatly appreciated. Funding was provided by the Social Science and Humanities Research Council of Canada, the Ontario-Québec Exchange Fellowships, the Ville de Québec, the CELAT (Université Laval), and the Fonds de soutien de revenue au doctorat, Département d'histoire, Université Laval. A large portion of my research was carried out at the Eastern Cereals and Oilseeds Research Centre (ECORC) at Agriculture Canada in Ottawa, a resource that was essential in the completion of this project, and is duly recognised. The BUGS database developed by Paul Buckland, Jon Sadler and Phil Buckland was also extremely helpful in gleaning out necessary ecological data on identified insect species. Others that helped with logistics, translations, identifications, and in so many other ways were; Bob Anderson, John Bain, Ed Becker, Klaus Bolte, Lynda Bottoms, Guylaine Boucher, Marie-Claude Boileau, Yves Bousquet, Don Bright, Céline Cloutier, Anthony Davies, Christiane Demers, Leslie Driscoll, Martine Duchaîne, Catherine Fortin, Henri Goulet, Manon Goyette, Bruce Gill, Vasily Grebennikov, Lynda Gullason, Darcelle Hall, Andrée Héroux, Anne Howden, Serge Laplante, Laurent LeSage, Jim McKenzie, Jeremy McNeil, Richard Morlan, William Moss, James O'Hara, Eva Panagiotakopulu, Steve Pendery, John Pinto, The Price Family (Québec City), Mike Schwartz, Bob Skidmore, Jim Smith, Brad Sweet, and Marie-Noëlle Tremblay. Members of the 1995 excavation team, not previously mentioned, who patiently helped gather my samples were Jacques-Alexis Bernadin, Frédéric Bussières, Philippe deVarennes, Anne Fournier, Stéphanie Heureux, Claude Lafleur, Sophie Lemelin, Magali Morlion, and Marc-Olivier Payant. Finally, this work is dedicated to my husband Anand.

APPENDIX A

FIGURES

Figure 1

Location of the Îlot Hunt Site (CeEt-110)

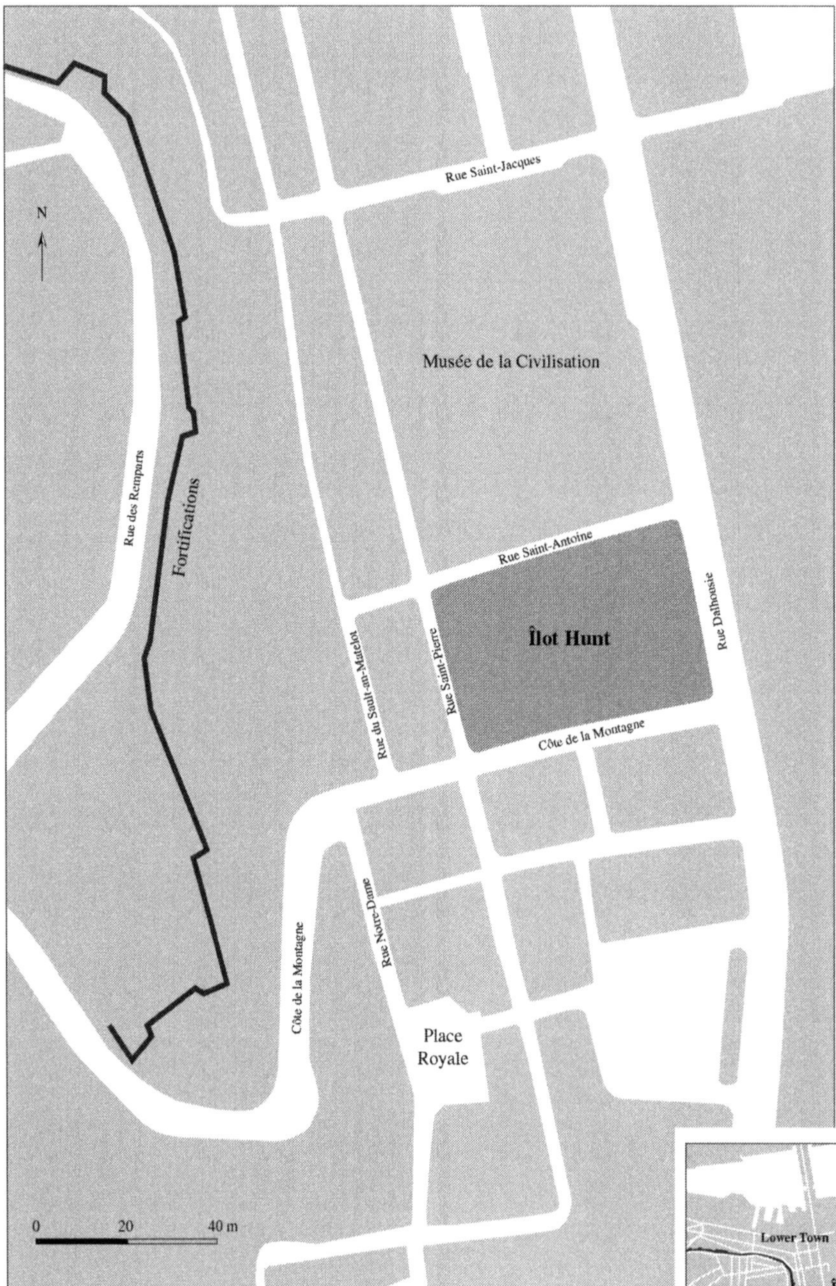

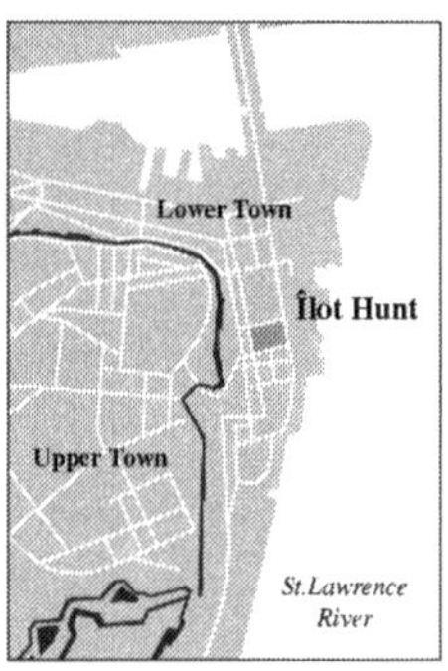

(Ville de Québec, Service de l'ingénierie, Division de l'arpentage et de la cartographie, Carte topographique au 1:1000, Feuillet 1124, 1993)

Figure 2

Plan of 16D200/16C26 privy and drains, Îlot Hunt Site

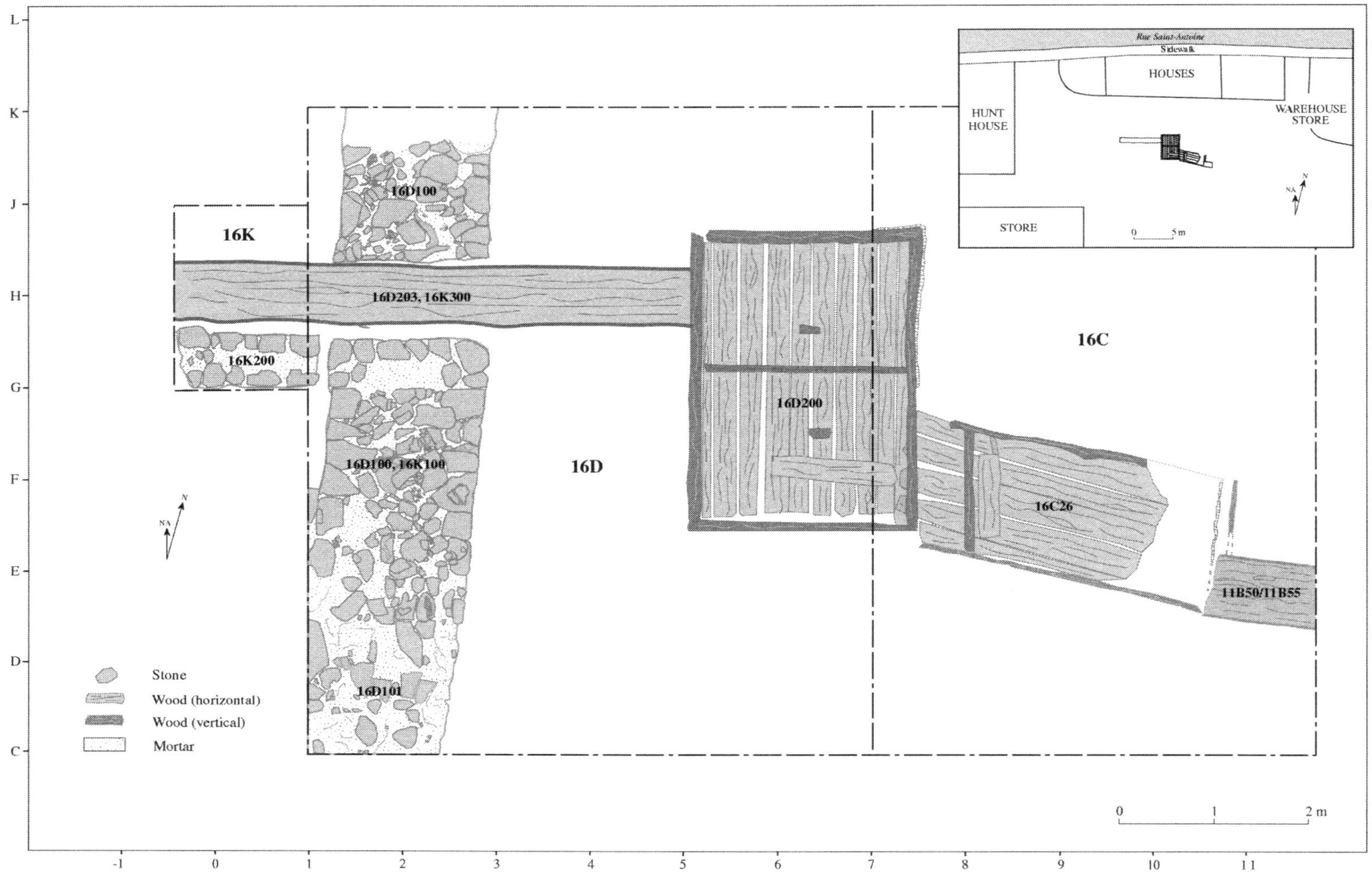

Figure 3

The Îlot Hunt Site in 1845

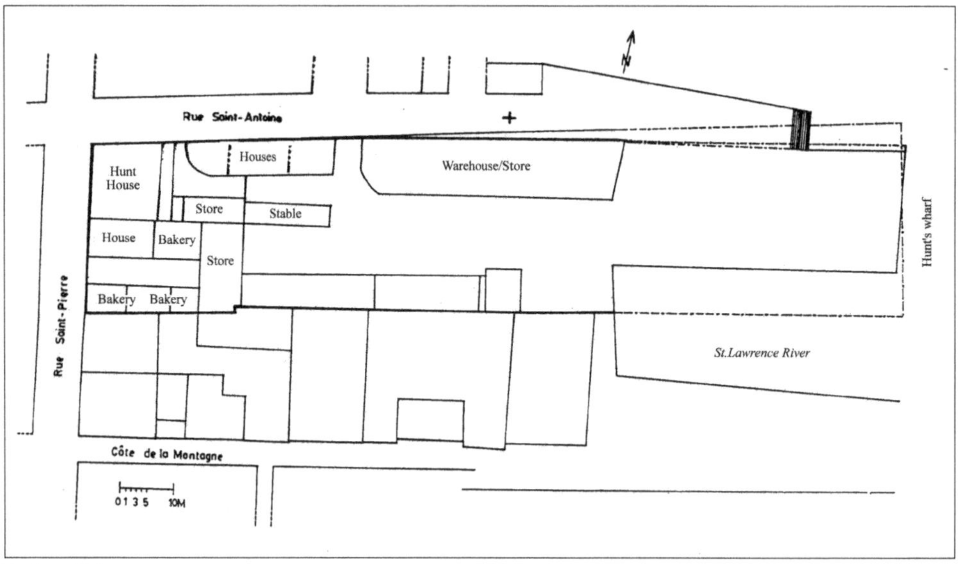

(Leclerc: 1998)

Figure 4

Sub-operation 11C and plan of 11C11 drain and 16D200 privy section

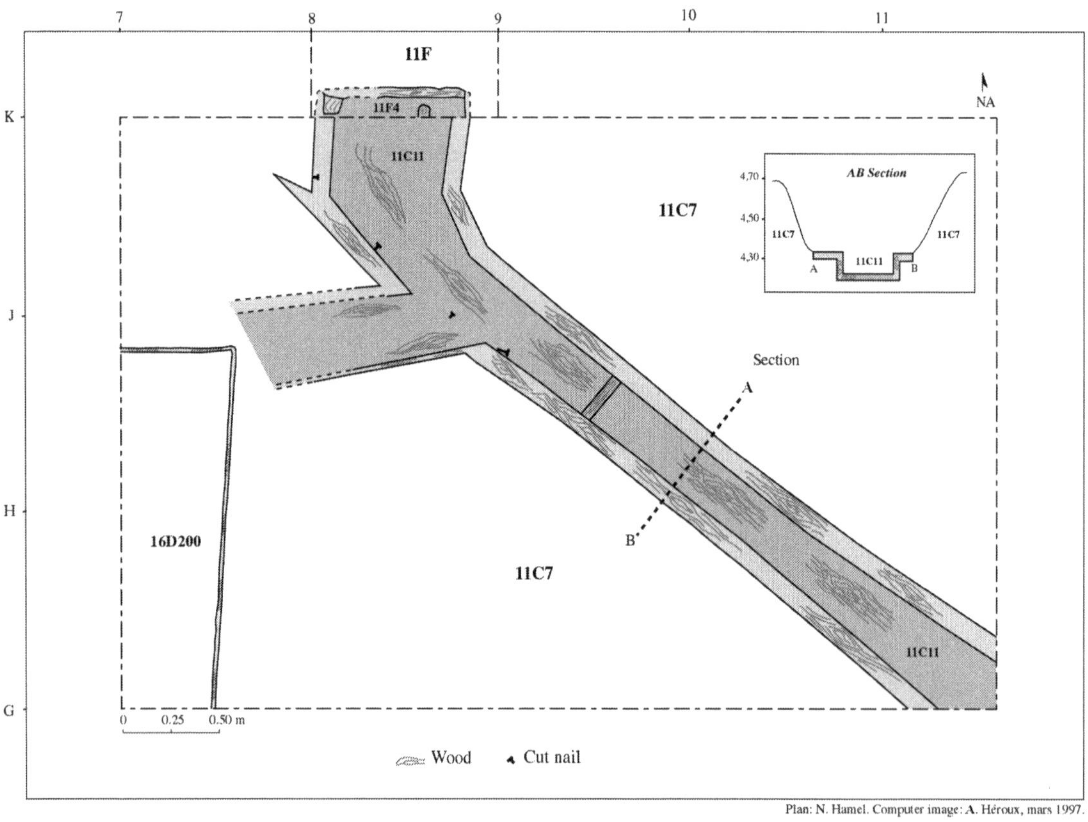

Figure 5

Québec City, in front of the Notre Dame Basilica in Upper Town, during the 1832 cholera epidemic

(«*Le choléra à Québec*», National Gallery of Canada, Ottawa, 1959 (7157), from David-Thierry Ruddel's, *Québec: 1765-1832*, 1991.)

Figure 6

Plan taken from 1879 Hopkins Atlas of Québec City indicating the city's wards

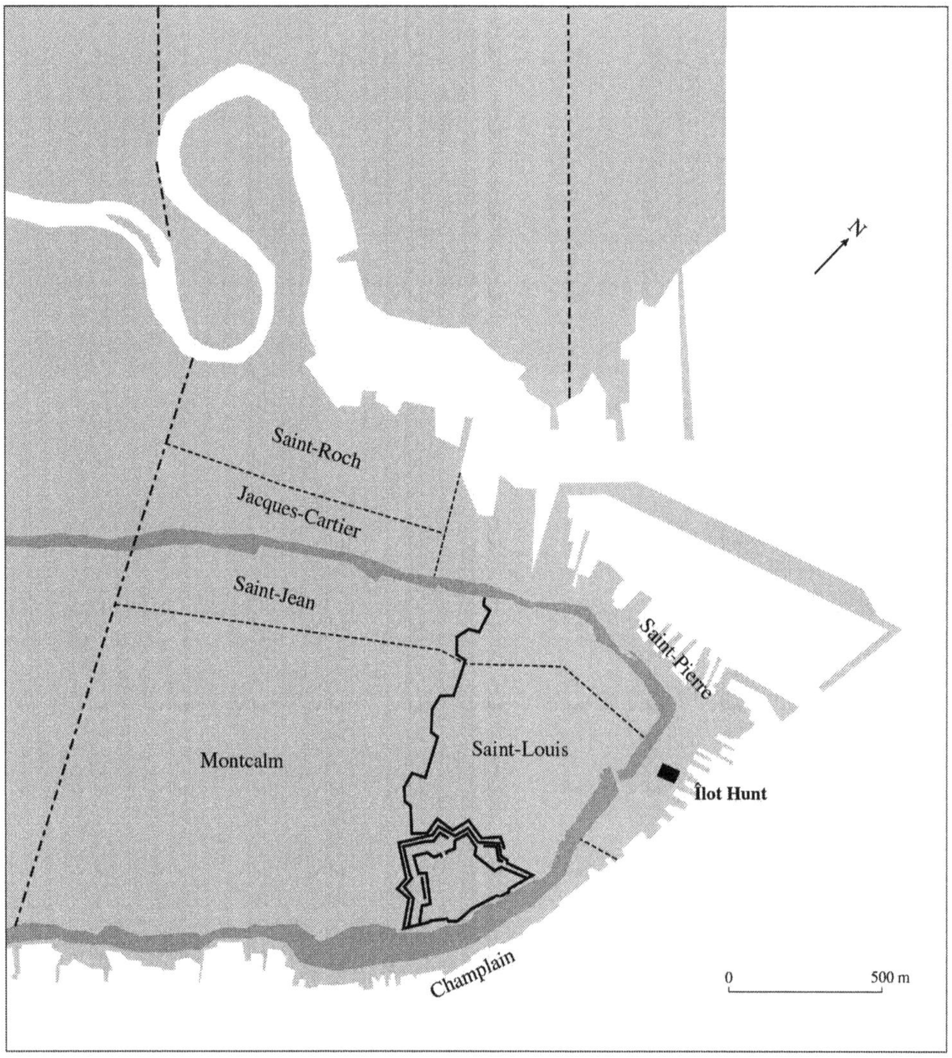

Figure 7

View of the port of Québec City in 1833

(«*This view of The Lower City of Quebec from the Parapet of the Upper City, Is by Special Permission dedicated to His Most Excellent Majesty William the Fourth. From a drawing by Lieut. Col. Cockburn. Aquatint by C. Hunt. London Pubd. 1833, by Ackerman & Co. 96*», *Strand.* Toronto, Metropolitan Toronto Central Library, Baldwin Room, J. Ross Robertson Collection (88). From Christina Cameron et Jean Trudel's, *Québec au temps de James Patterson Cockburn*, 1976.)

Figure 8

Cadastral lots of the Îlot Hunt Site

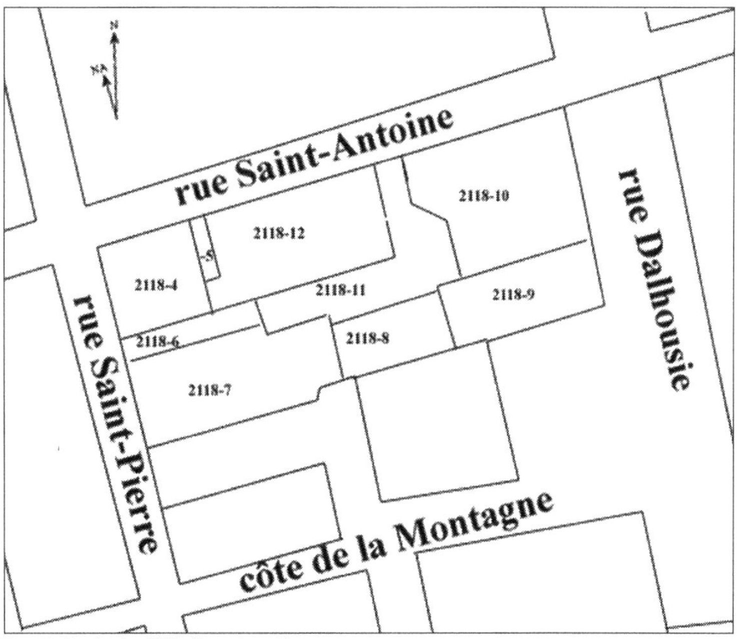

(From Laframbroise 1991: figure 1, p. 7)

Figure 9

Plan of Îlot Hunt Site in 1744

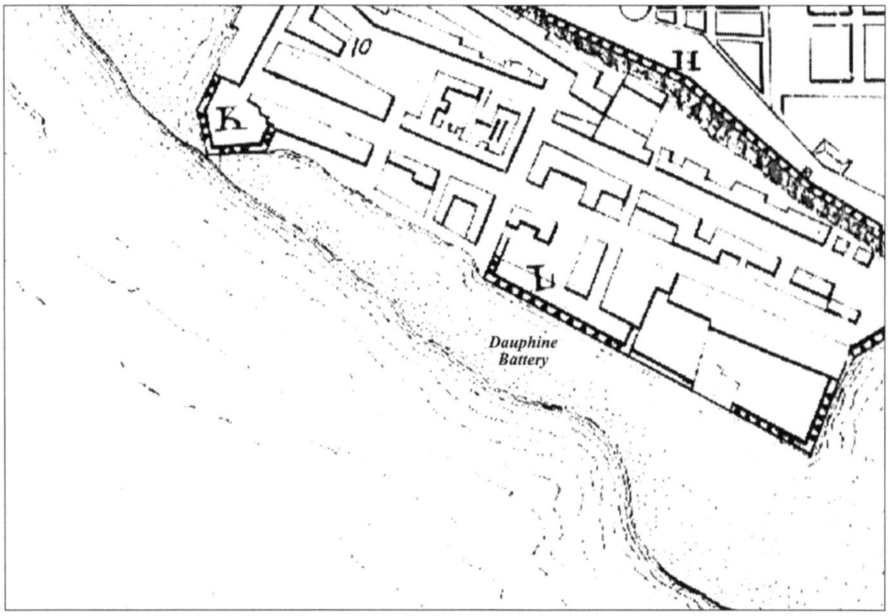

(National Archives of Canada, H3/340-Québec-ca1744, *Plan de la ville de Québec*)

Figure 10

Plan of Îlot Hunt in 1815 indicating construction of a house on rue Saint-Antoine

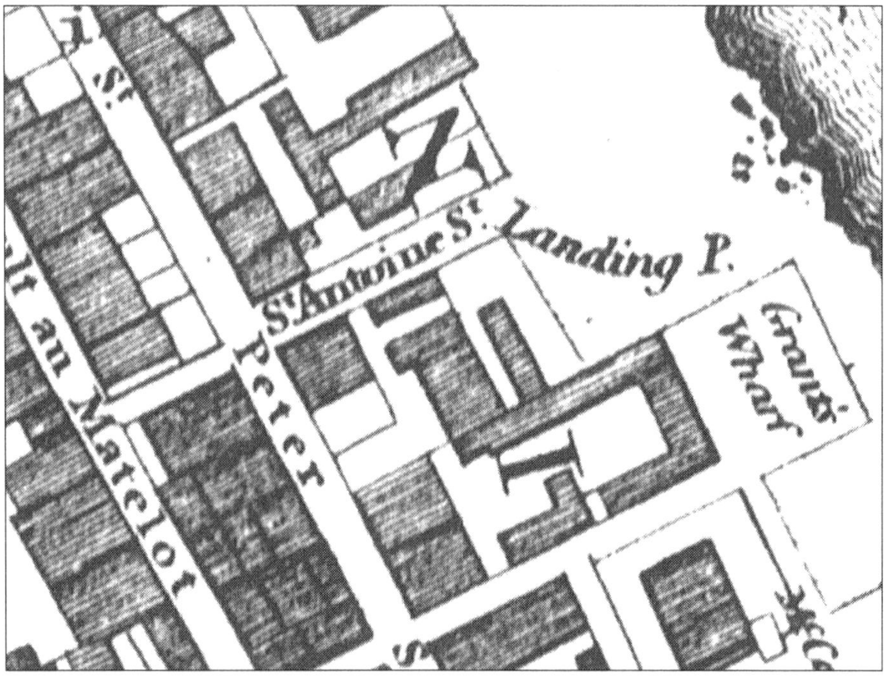

(Archives nationales du Québec-Québec, Joseph Bouchette, *Carte topographique de la Province du Bas-Canada*)

Figure 11

Southeastern view along rue Saint-Antoine showing houses dating to c.1815 and c.1824

(*Québec, rue Saint-Antoine, 1952-05-02*, Archives nationales du Québec-Québec, Fonds Inventaires des œuvres d'art, E6, S8 P11464-A-1.)

(c.1815 house is closest to camera.)

Figure 12

1824 plan showing second house on rue Saint-Antoine

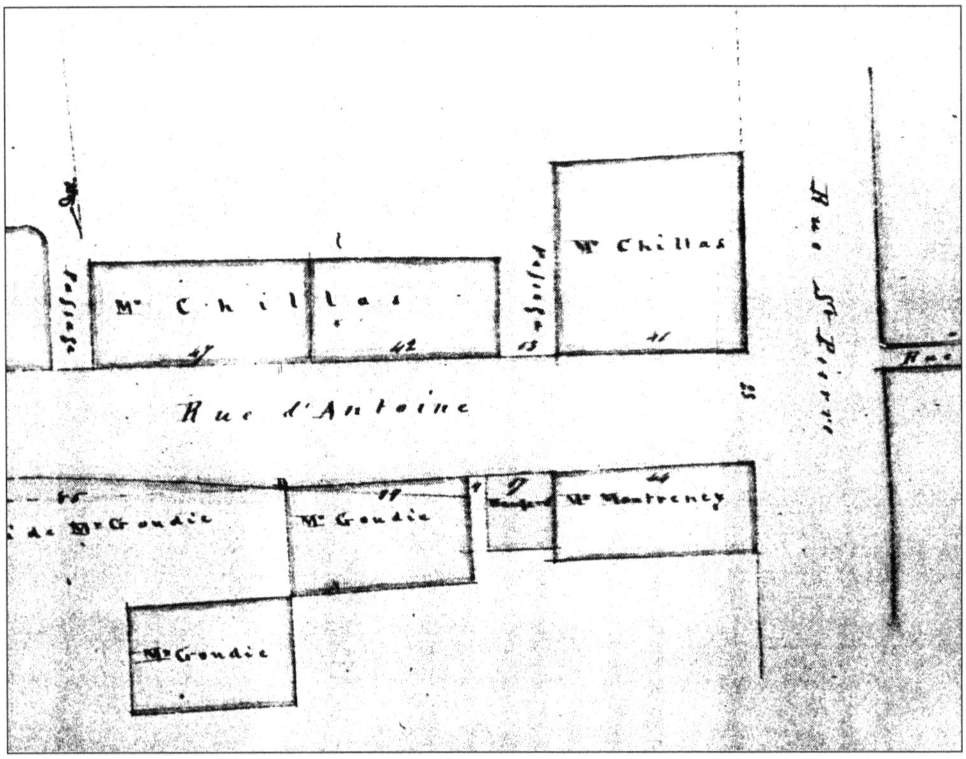

(Archives de la ville de Québec)

Figure 13

Îlot Hunt in 1824

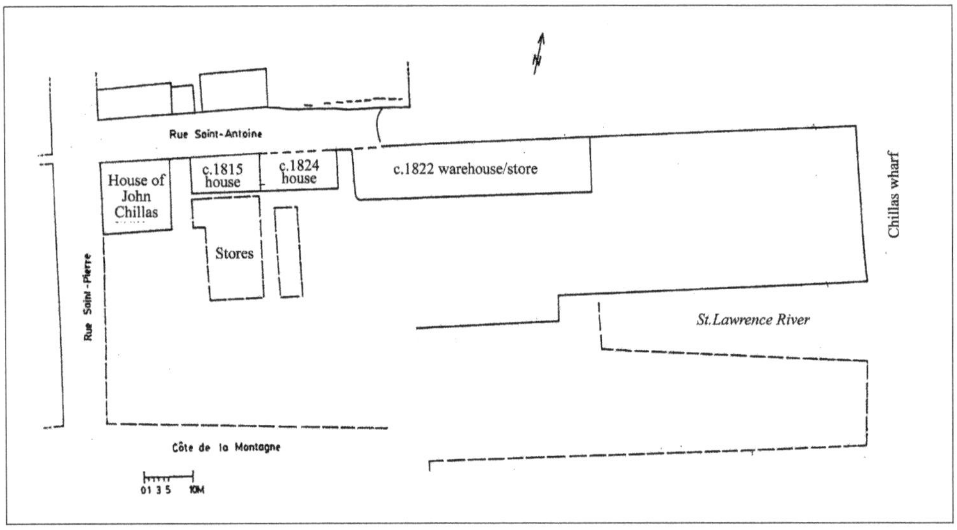

(Leclerc: 1998)

Figure 14

Jean-Baptiste Larue's plan of Îlot Hunt and proposed extension of rue Saint-Antoine in 1829

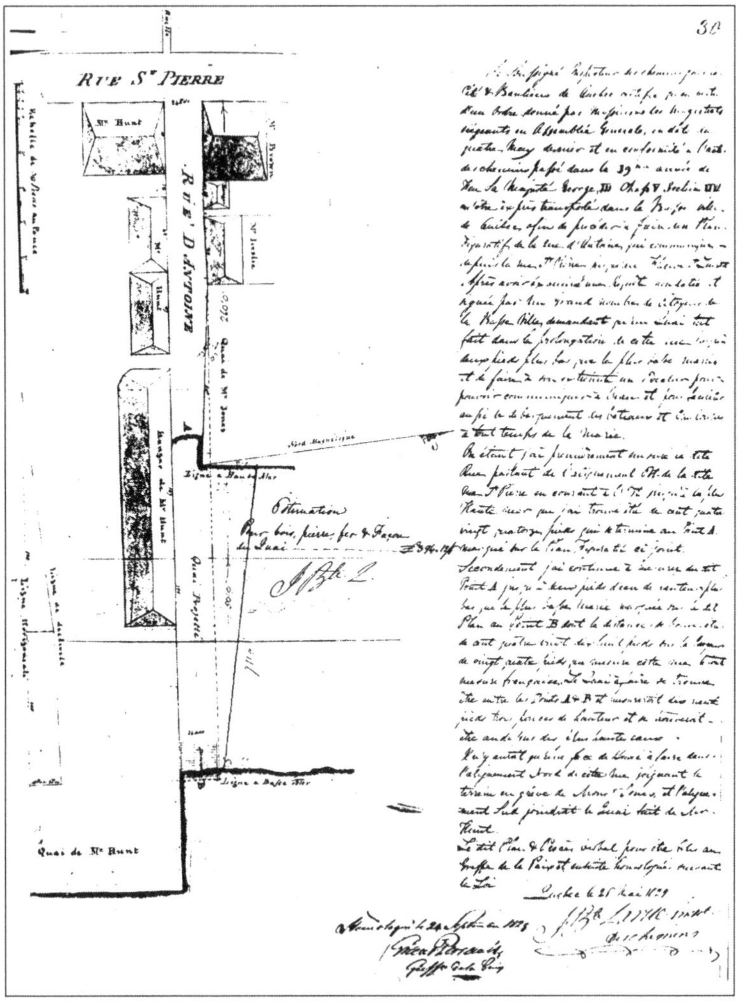

(*Plan figuratif de la rue d'Antoine, 1829.* Jean-Baptiste Larue, A.V.Q., D-344, 12-1829)

Figure 15

1845 plan of Îlot Hunt

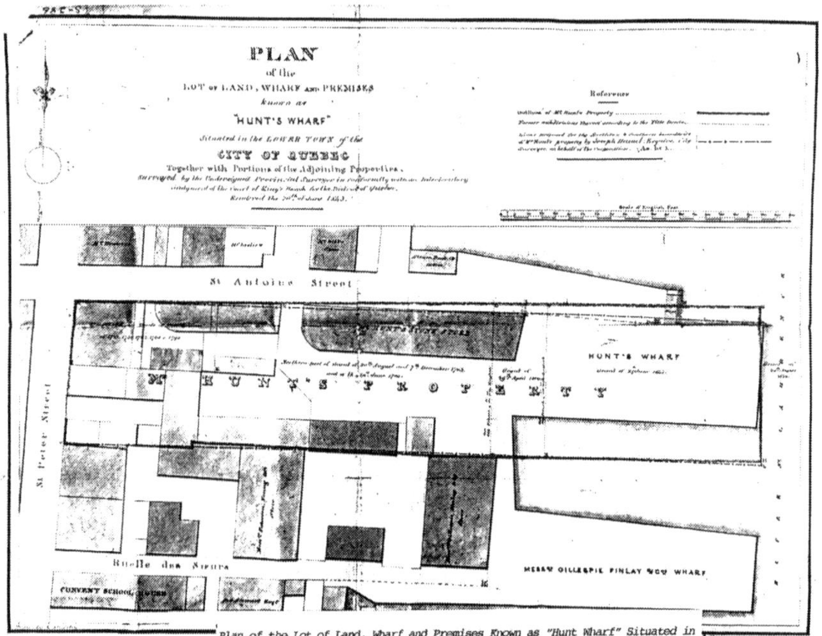

Plan of the Lot of Land, Wharf and Premises Known as "Hunt Wharf" Situated in the Lower Town of the City of Quebec Together with Portions of the Adjoining Properties, 20 juin 1843 (24 mars 1845?). A.V.Q. B-344.12-1845

(Note rounded corners on c.1815 house and warehouse/store.)

Figure 16

Îlot Hunt in 1875

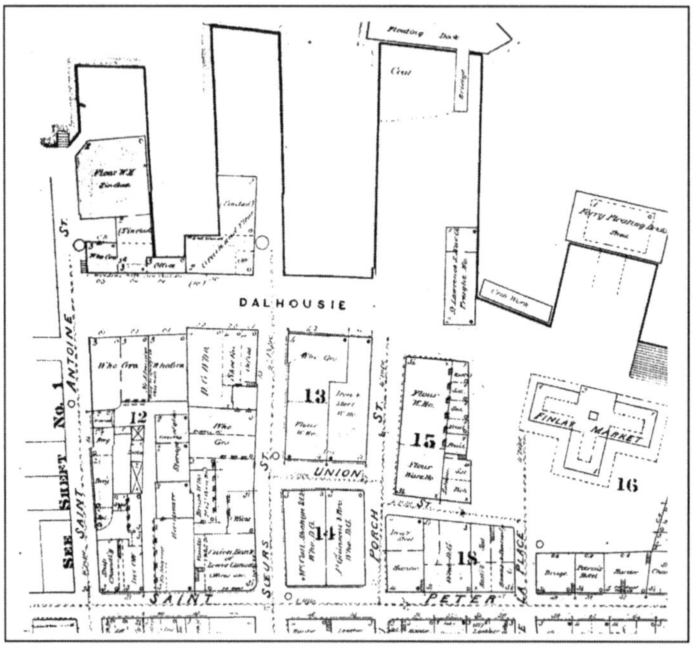

Insurance Plan Of the City of Quebec, Canada. Published by D.A. Sanborn in Broadary N.Y. Atlas Sanborn fo.2, nég. no: N1173-11. A.V.Q. B-342-1875

Figure 17

Plan of îlot Hunt excavations (1988, 1991, 1992, 1993, 1994, 1995)

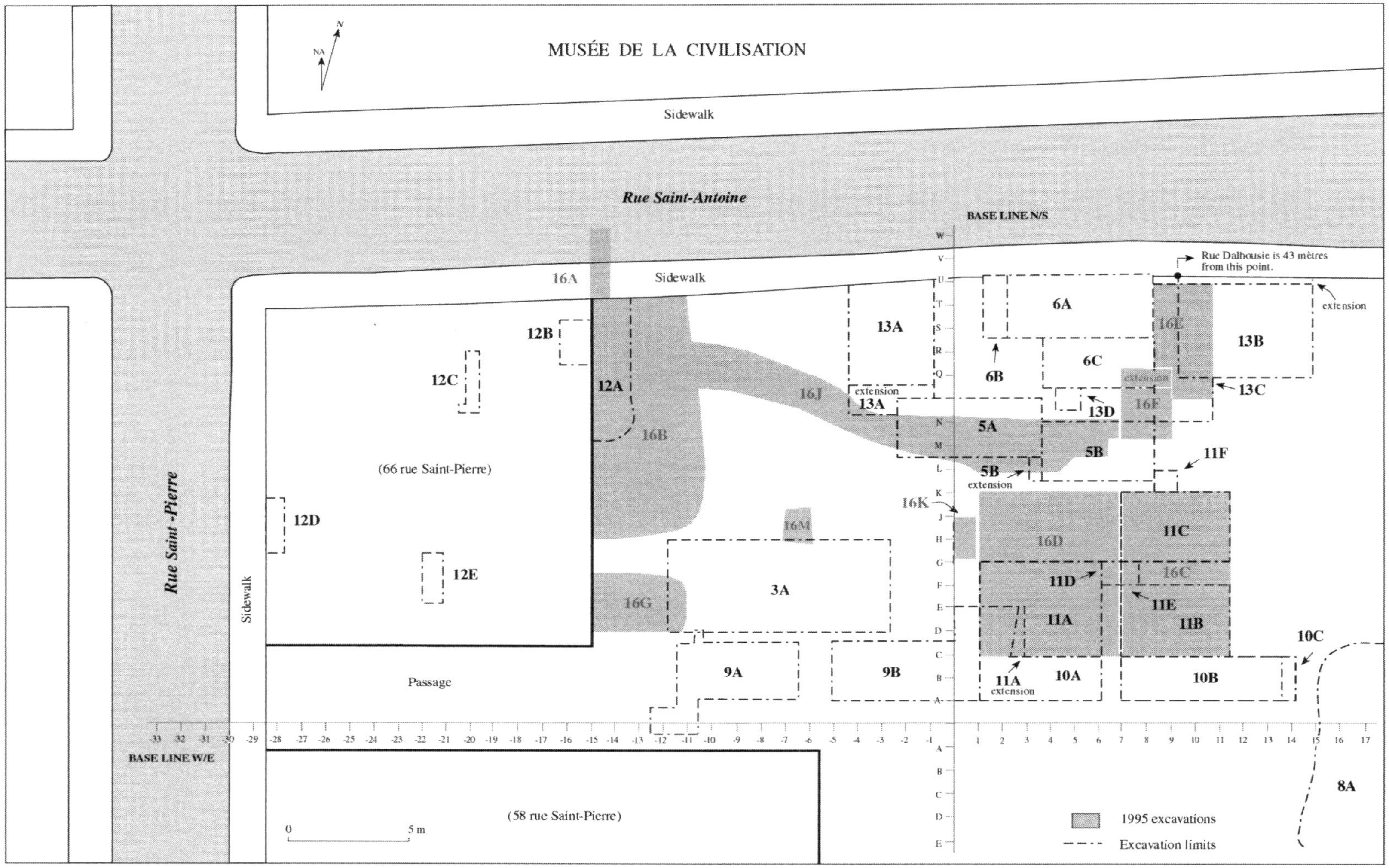

Figure 18

Îlot Hunt, CeEt-110
Plan of excavated structures (1988, 1991, 1992, 1993, 1994, 1995)

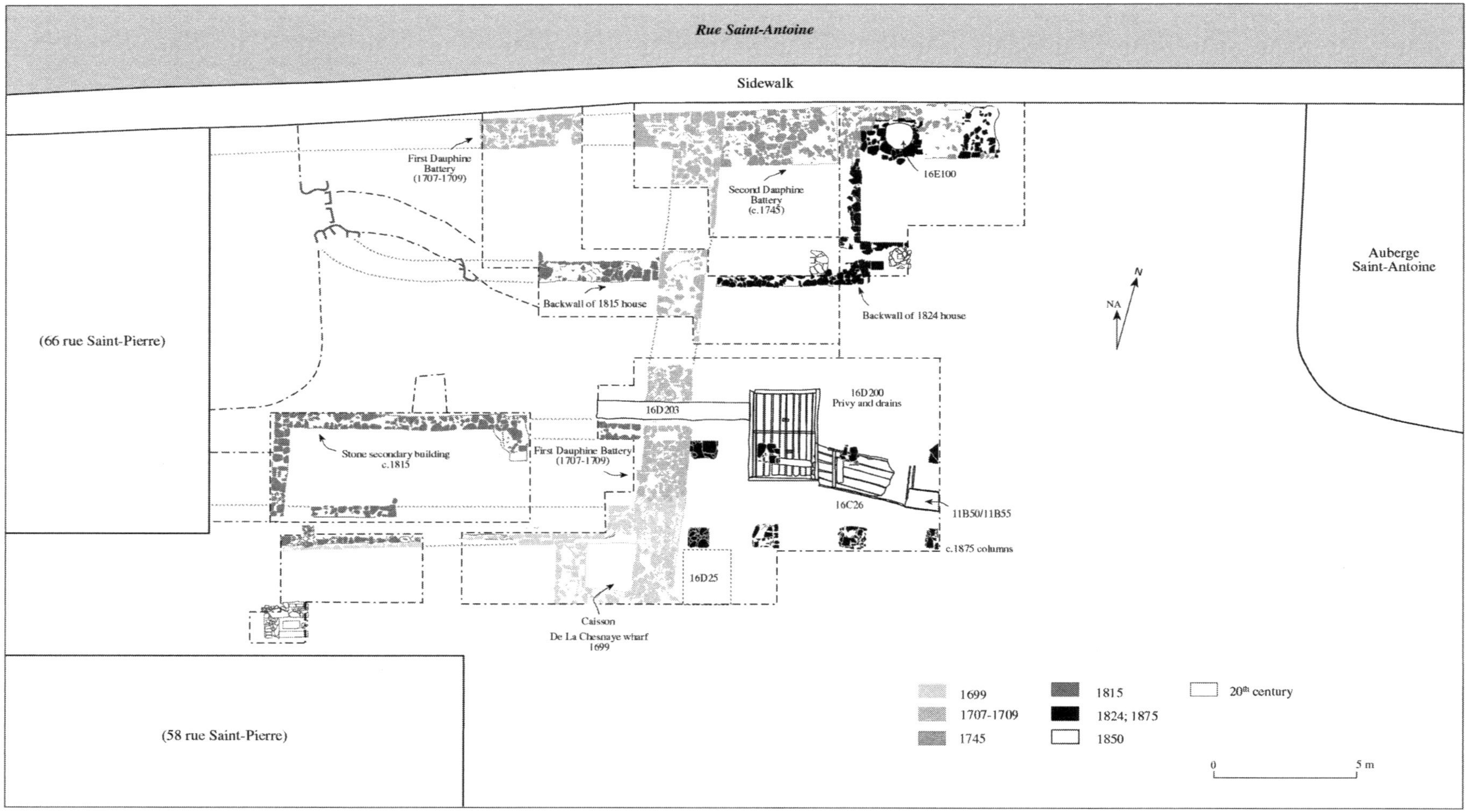

Figure 19

11B50/11B55 covered drain at eastern end of 16C26 privy section

(Photo: Laboratoire d'archéologie historique, Université Laval)

Figure 20

Îlot Hunt, CeEt-110

Sub-operations 16C and 16D, First Dauphine Battery (1707-1709) and columns from 1875 structure

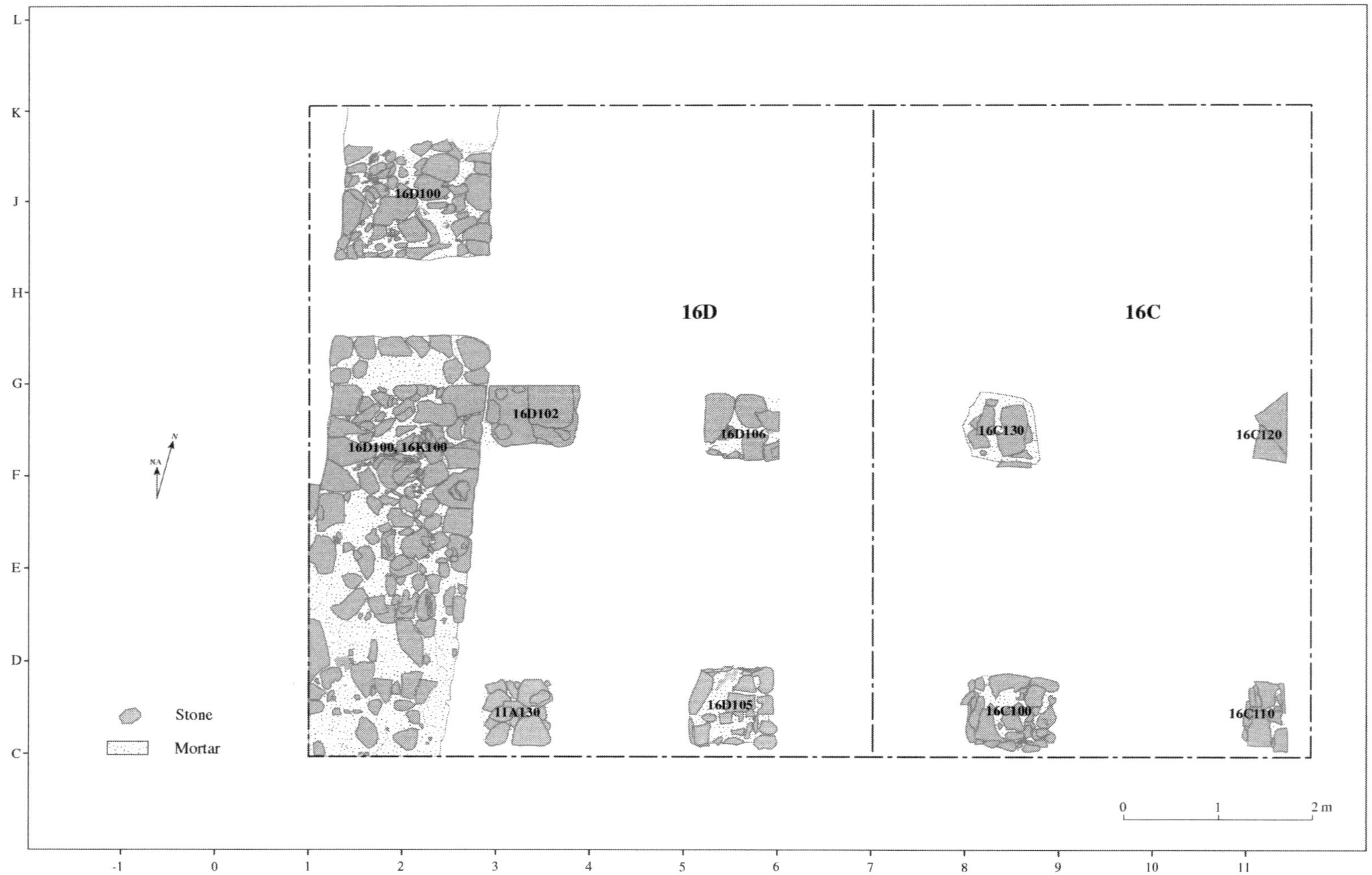

Figure 21

16C26 privy section

(Negative #95NB8-26)

Figure 22

Location of level 16C74 in 16C26 privy section

(Negative #95D14-19)

Figure 23

Eastern edge of de la Chesnaye wharf (16D101) and First Dauphine Battery (16D100)

(Negative #95NB9-8)

(Note: excavation to c.1707 beach level in front of structure.)

Figure 24

Profile of 16D25 in southern wall of 16D sub-operation

(Negative #95NB9-2)

Figure 25

Sub-operation 16D, interior of 16D200 privy section
Stratigraphic profile (facing west) taken from north-south berm

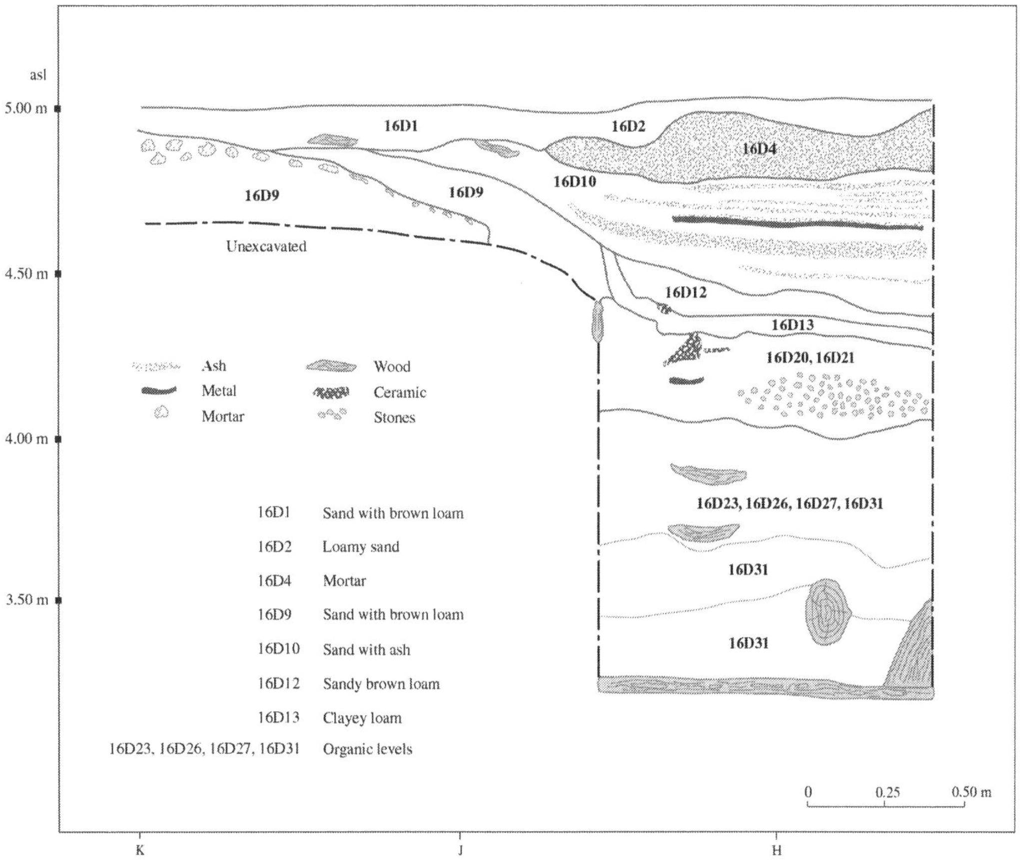

Figure 26

16D200 section of privy, southwards view

(Negative #95D10-20)

End of excavation of layer 16D26 (on left) and beginning of layer 16D24 (on right, with wood).

Figure 27

View (facing west) of northeastern corner of 11C15/11C35/16D200 privy section

(Photo: Laboratoire d'archéologie historique, Université Laval)

Figure 28

Sub-operation 11C, containing privy structure (11C15/11C35/16D200) Stratigraphic profile from western edge of 1993 excavations

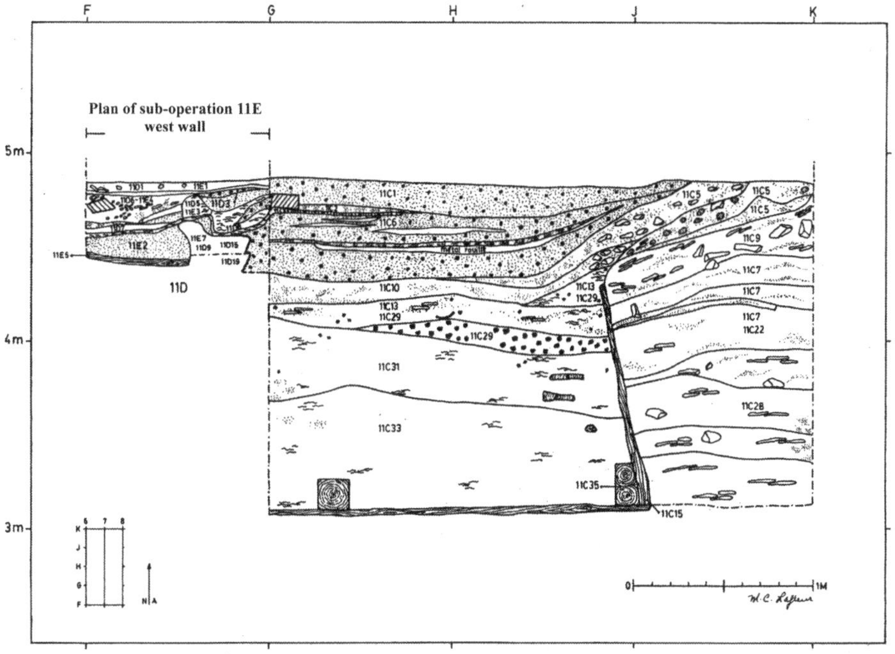

Figure 29

Flown blue plate marked "Daly's Hotel Montreal"

(Photo: Laboratoire d'archéologie historique, Université Laval)

Figure 30

Interior of 16D200 privy section with 16D106 column
Stratigraphic profile of southern surface of east-west berm

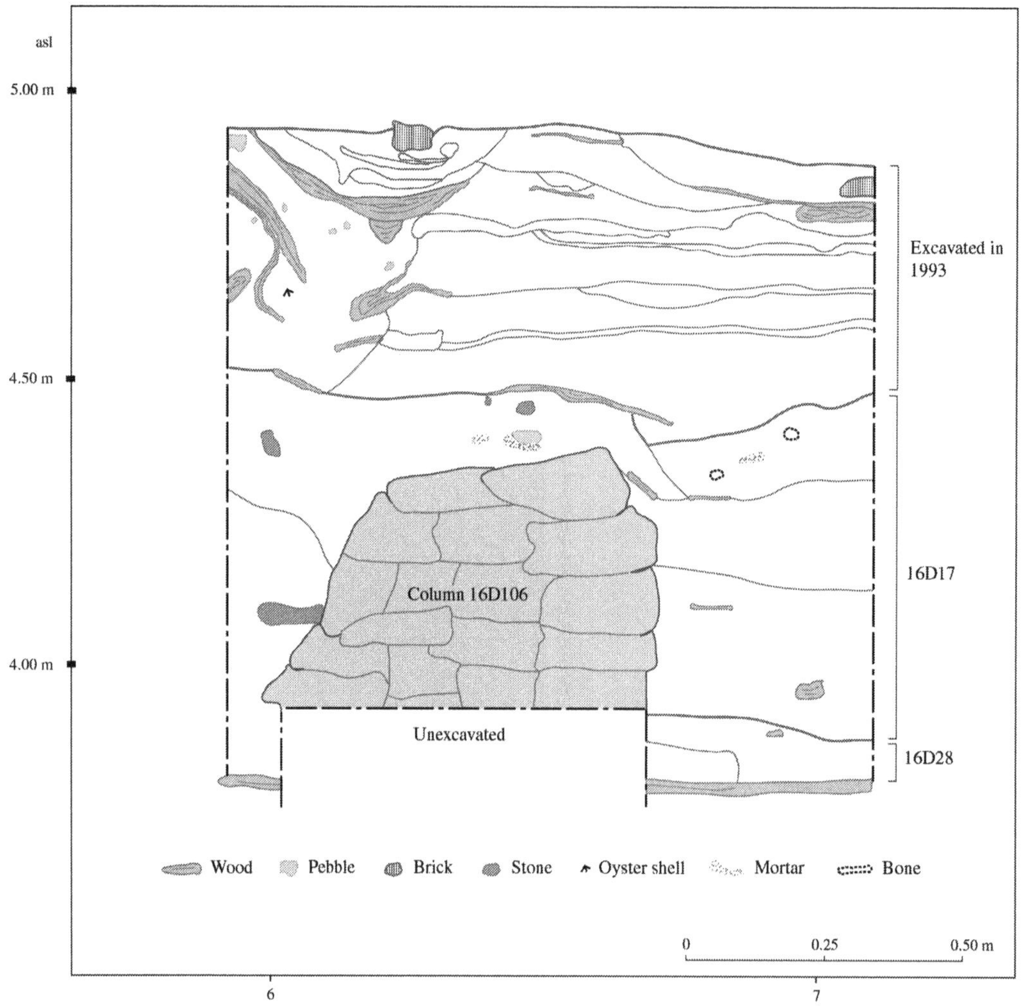

Figure 31

Interior of 16D203/16K300 drain

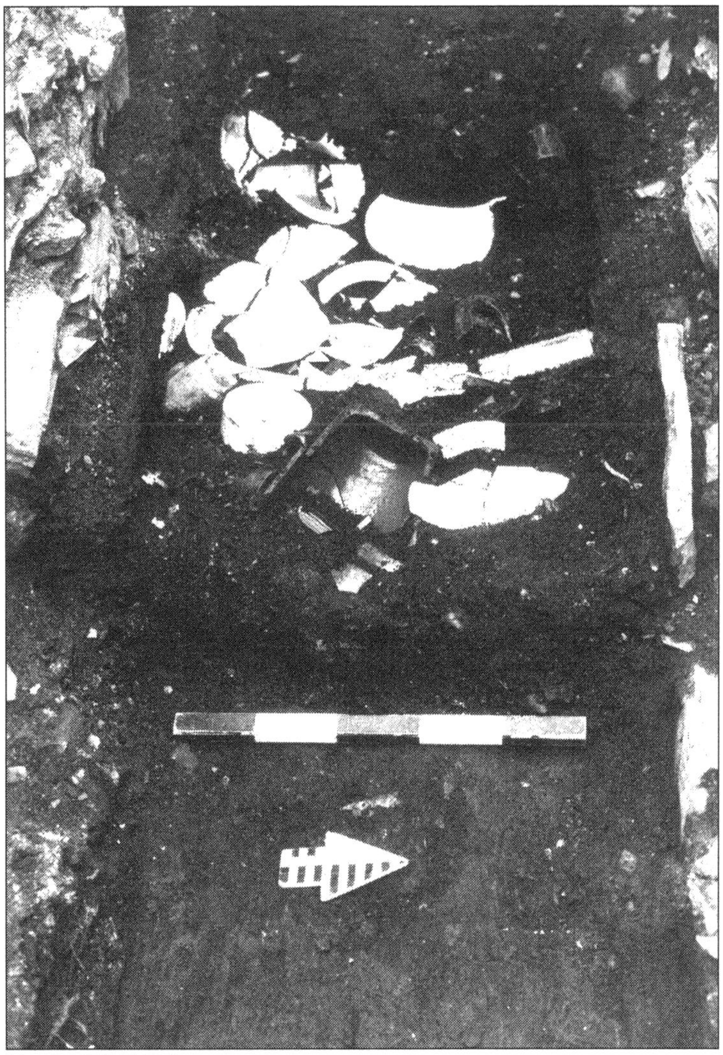

(Negative #95NB10-26)

Figure 32

Destruction of First Dauphine Battery (16D100) to construct 16D203/16K300 drain

(Negative #95NB3-14)

Figure 33

Sub-operation 16K, stratigraphic view of the west wall

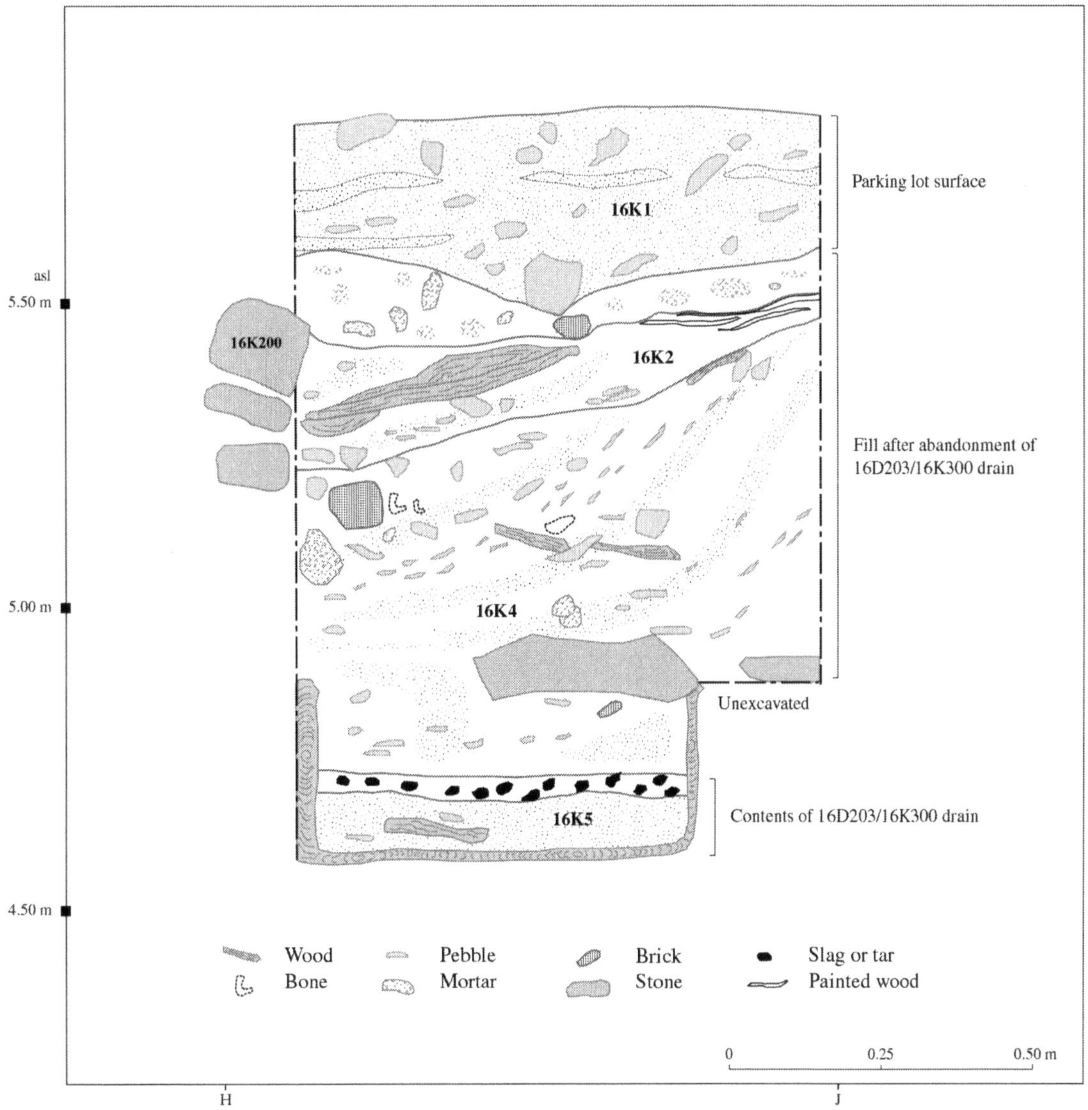

Figure 34

Sub-operations 16E and 16F

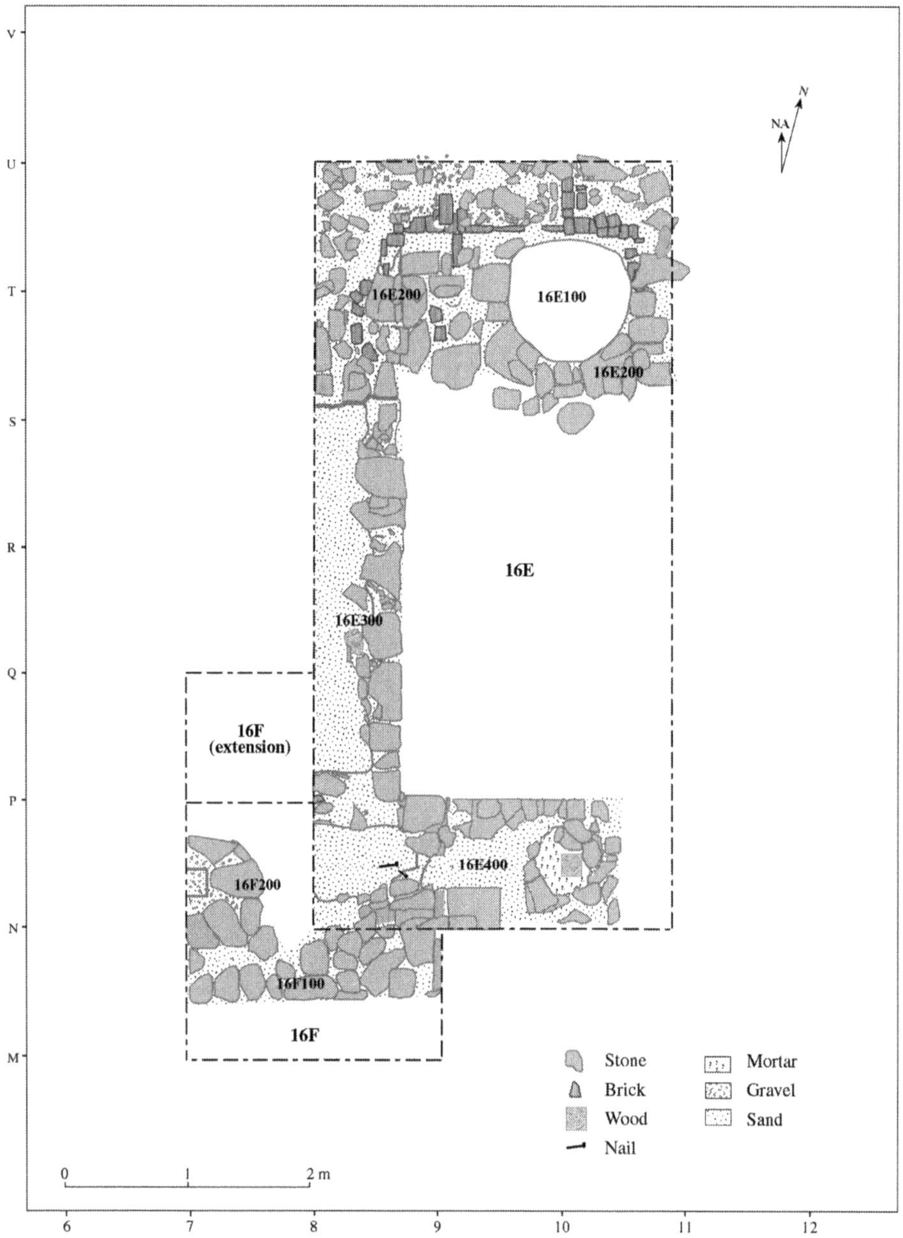

Figure 35

Post-excavation infilling of 16E100

Figure 36

Îlot Hunt, CeEt-110
Structures related to sanitary practices c.1850

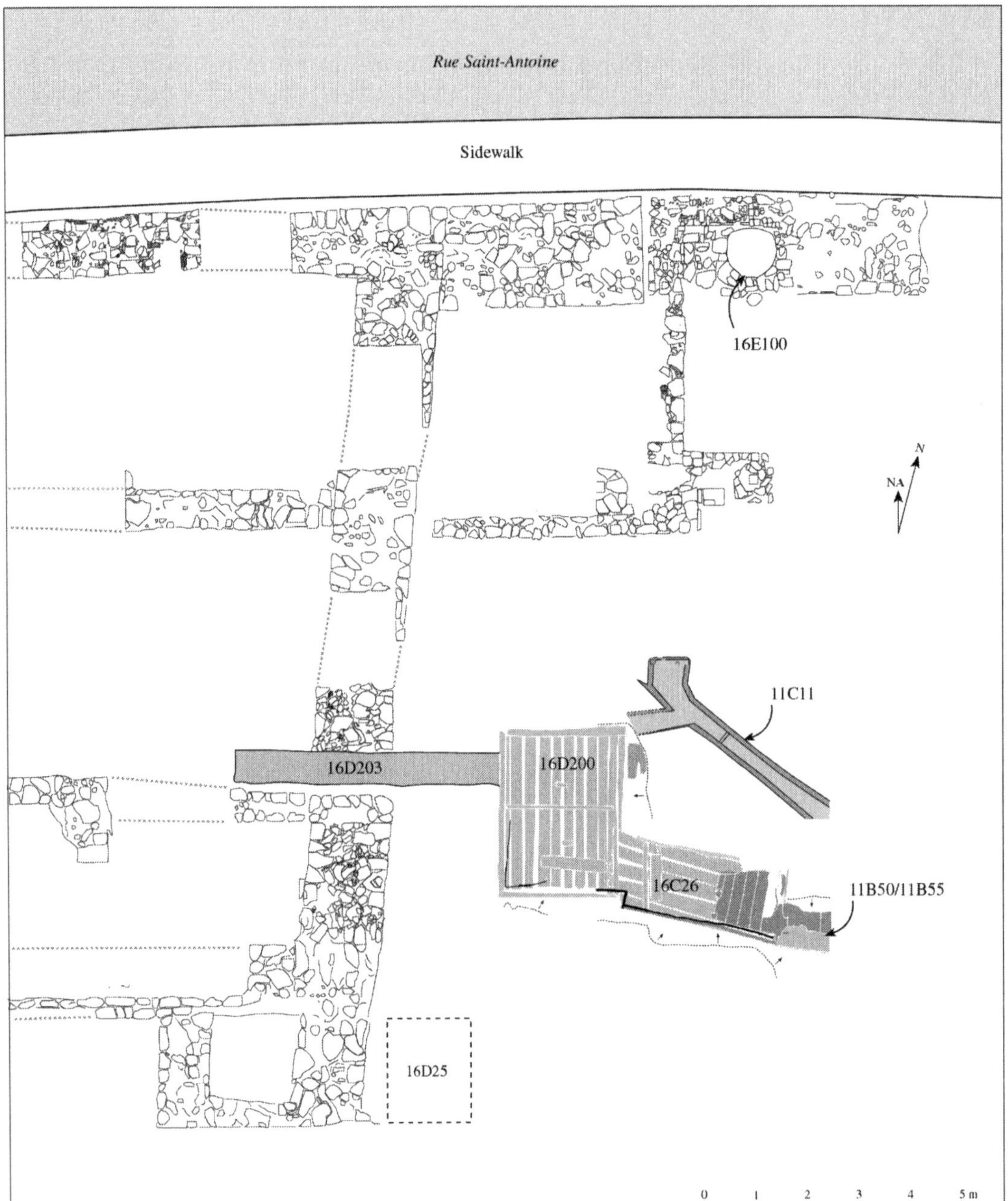

Figure 37

Îlot Hunt, CeEt-110
Structures related to sanitary practices c.1860-1875

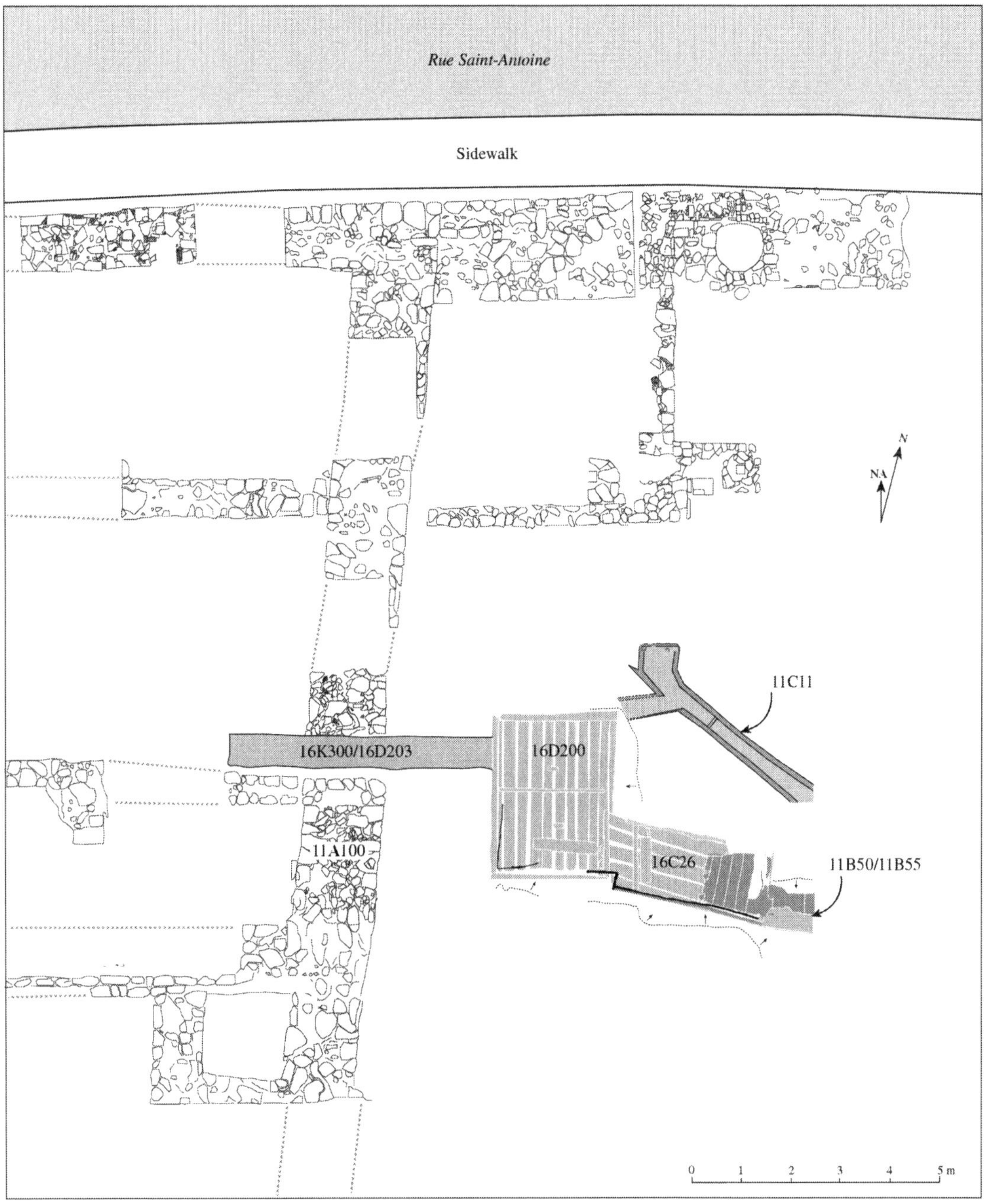

Figure 38

Transfer print pitcher from level 16D17

Figure 39

Brushes and toothbrushes from îlot Hunt

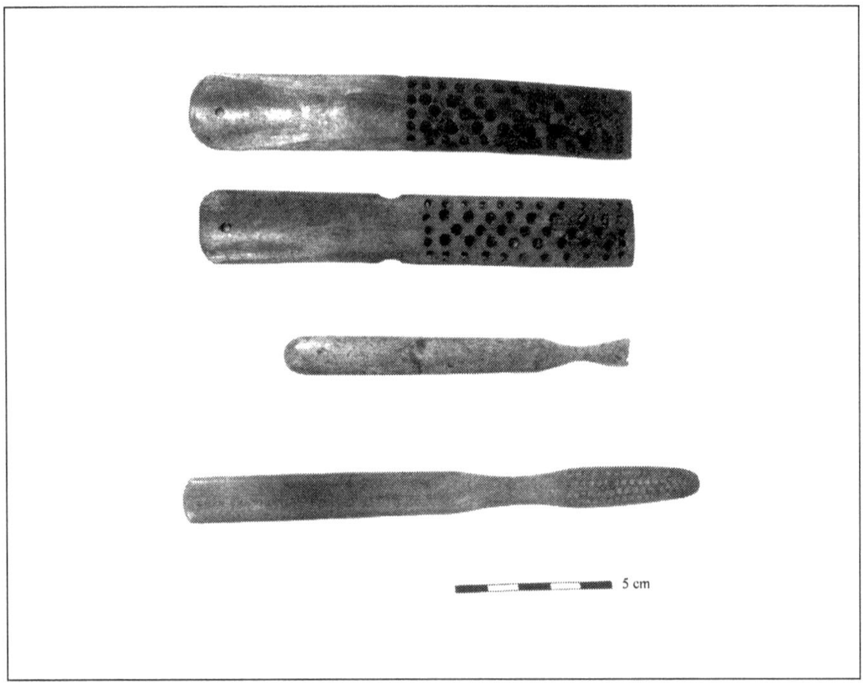

(Top 16C53-3 c.1870
Secon from top 16D41-19 c.1860
Second from botton 16D18-8 c.1880-1900
Bottom 16C71-4 c.1870)

Figure 40

Profile view of brush 16C53-1 dating to c. 1870

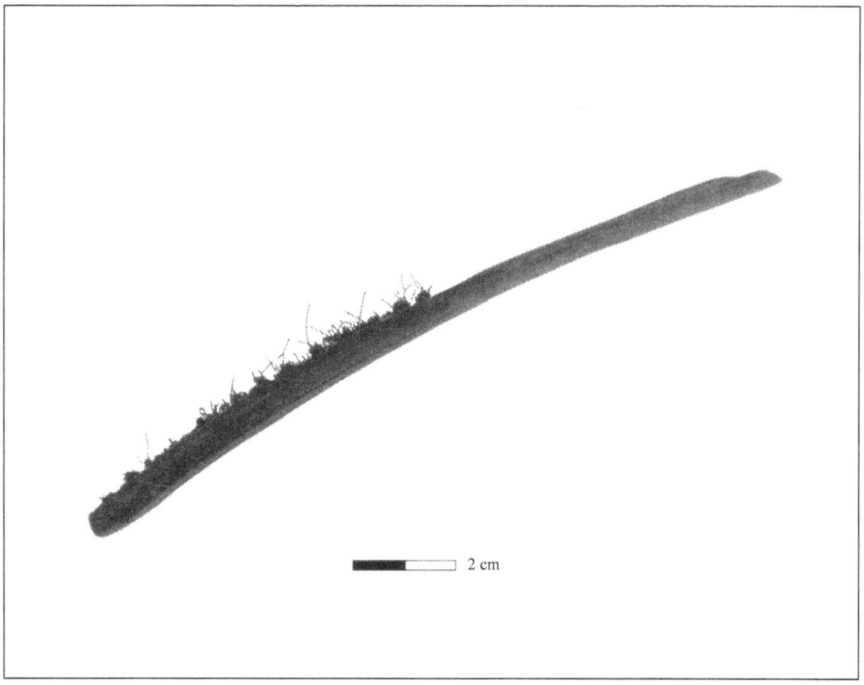

(97D7-1, L. Jodoin)

Figure 41

Medicinal bottle 16C53-2 dating to c.1870

(97D8-19, L. Jodoin)

Figure 42

Medicinal bottle 16D17-128 dating to c.1875

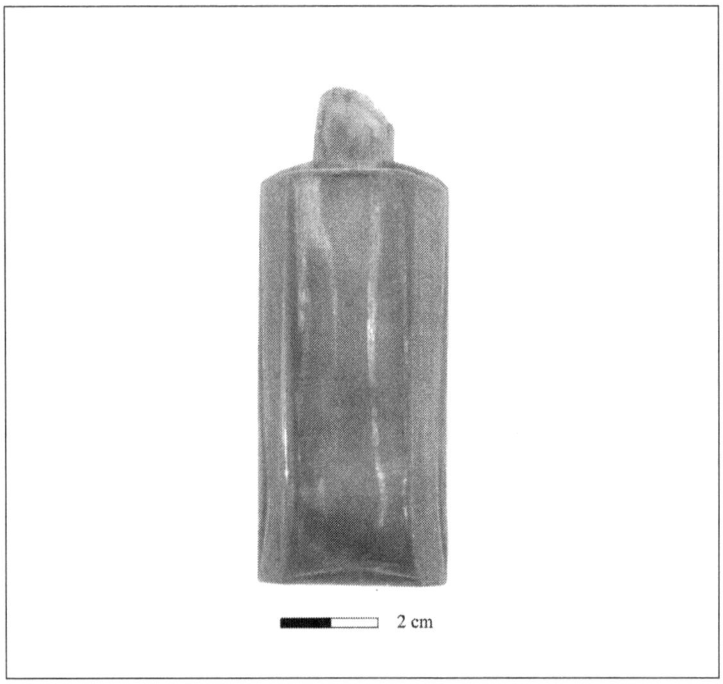

(98D2-13, L. Jodoin)

Figure 43

Vial 16C70-3 dating to c.1860

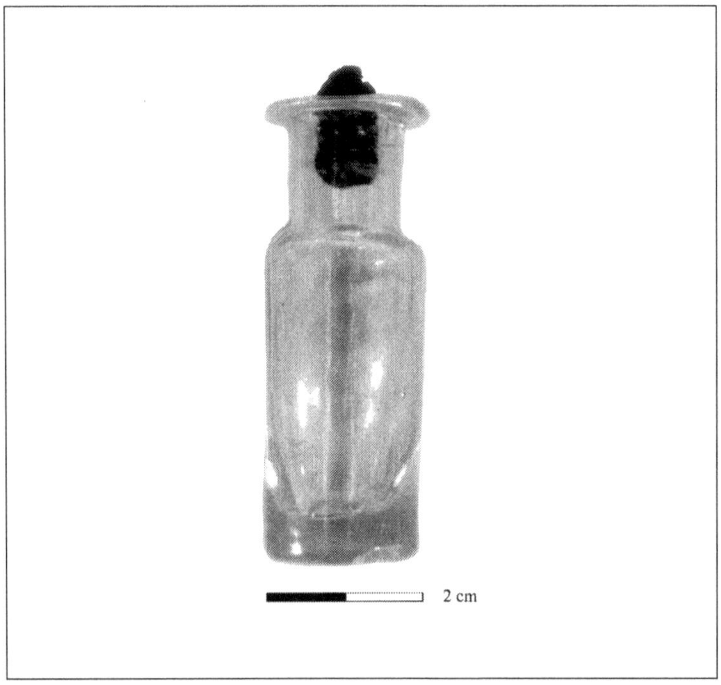

(97D8-12, L. Jodoin)

Figure 44

Vial 16D35-2 dating to c.1860

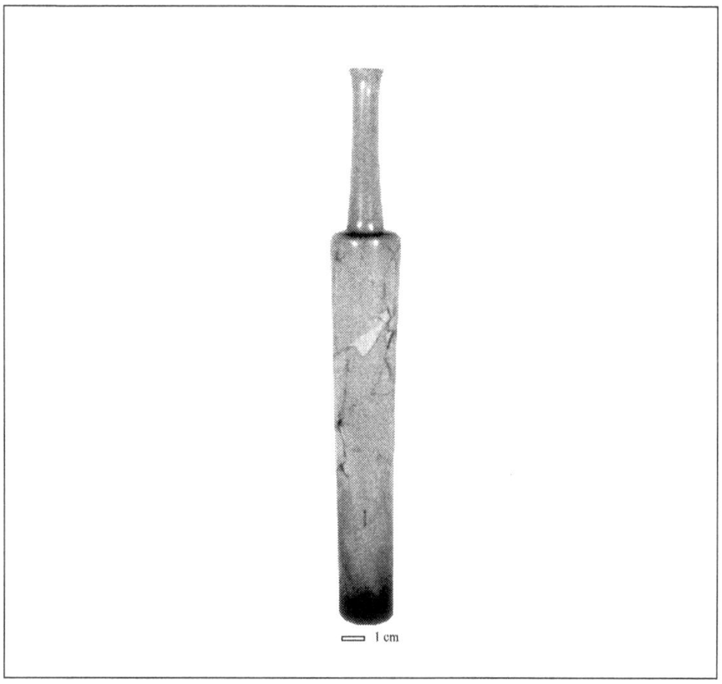

(97D3-12, L. Jodoin)

head of *Anotylus rugosus (Fabricius)*
(16C74)

45a .025 cm

pronotum of *Nausibius clavicornis* (Kugelmann)
(11C33)

45b .05 cm

head & pronotum of *Sitophilus granarius* (Linne)
(16D31)

45c .08 cm

left elytron of *Polygraphus rufipennis* (Kirby)
(16C74)

45d .056 cm

pronotum of *Bembidion tetracolum* Say
(16C74)

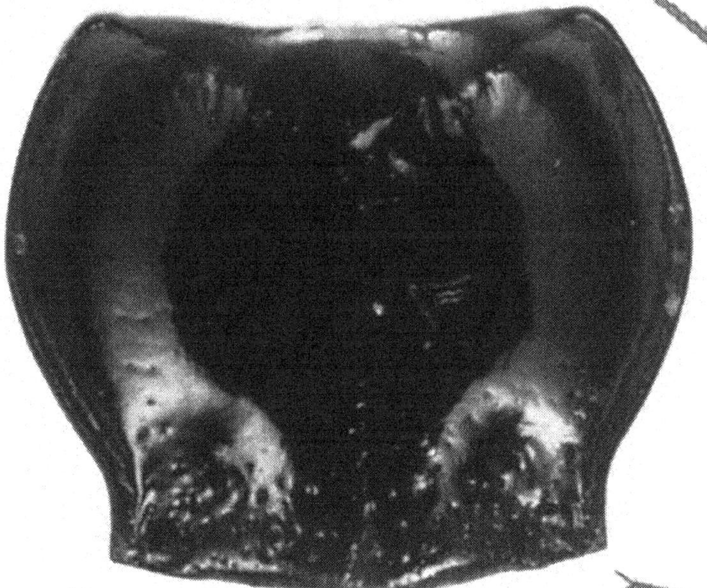

46a

.044 cm

Blatella germanica (Linne)
(Canadian National Collection)

46b

.31 cm

pronotum of *Blatella germanica* (Linne)
(16C66)

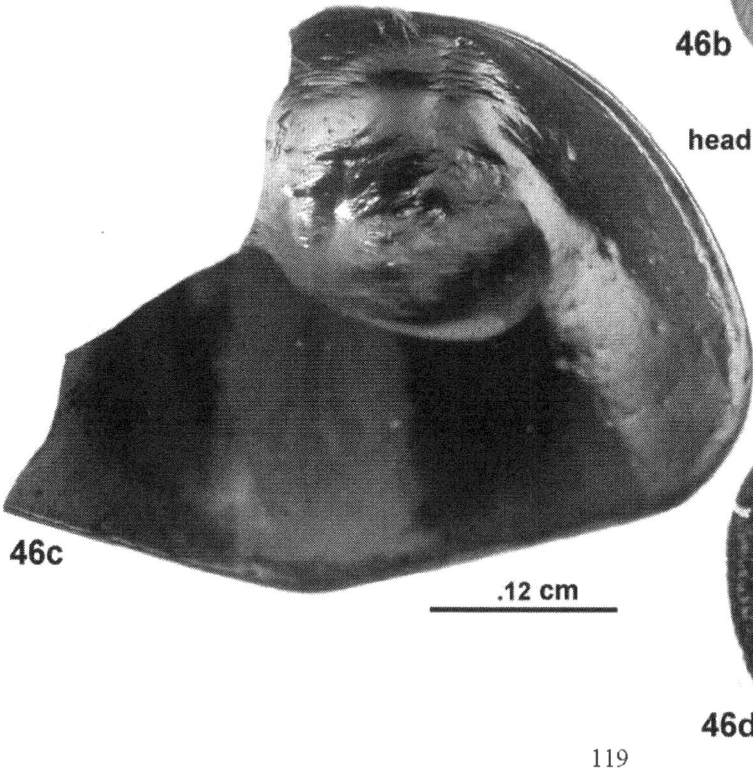

46c

.12 cm

head & pronotum of *Cimex* cf. *lectularis* Linne
(11C33)

46d

.05 cm

right elytron of *Trox scaber* (Linne)
(16D31)

47a

head of *Dryophthorus americanus* Bedel
(16E4)

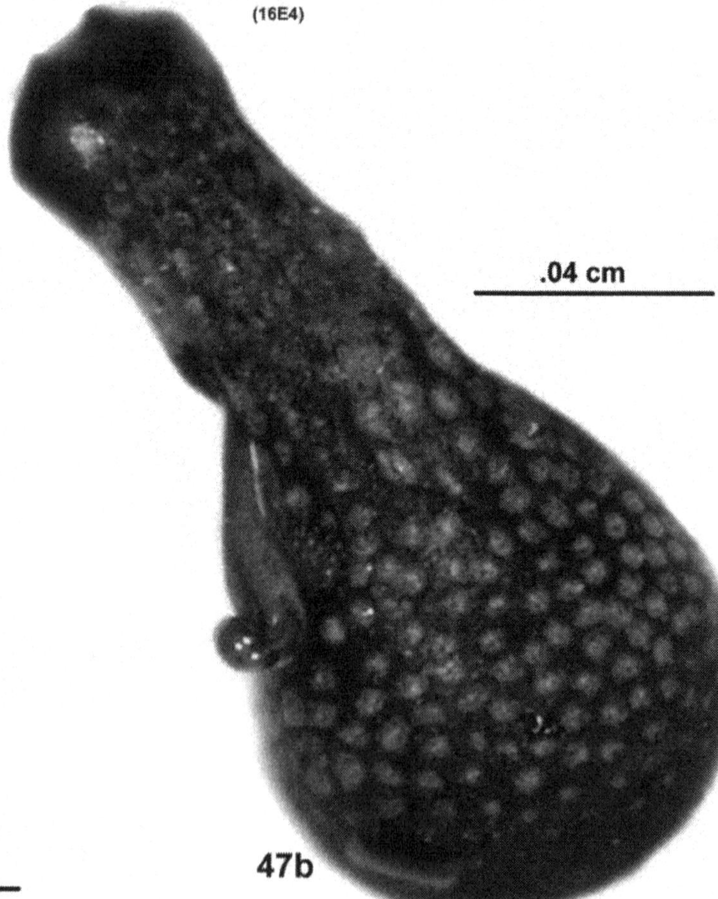

47b

left elytron of *Necrodes surimanensis* (Fabricius)
(11C33)

pronotum of *Creophilus maxillosus* (Linne)
(11C33)

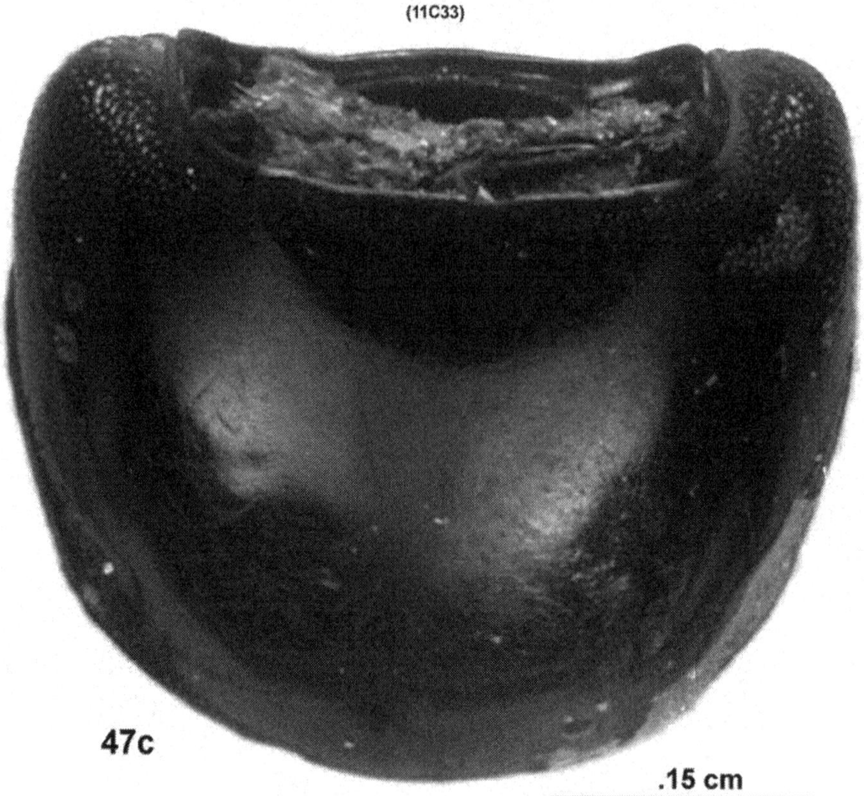

47c

47d

right elytron of *Omosita colon* (Linne)

(16C65)

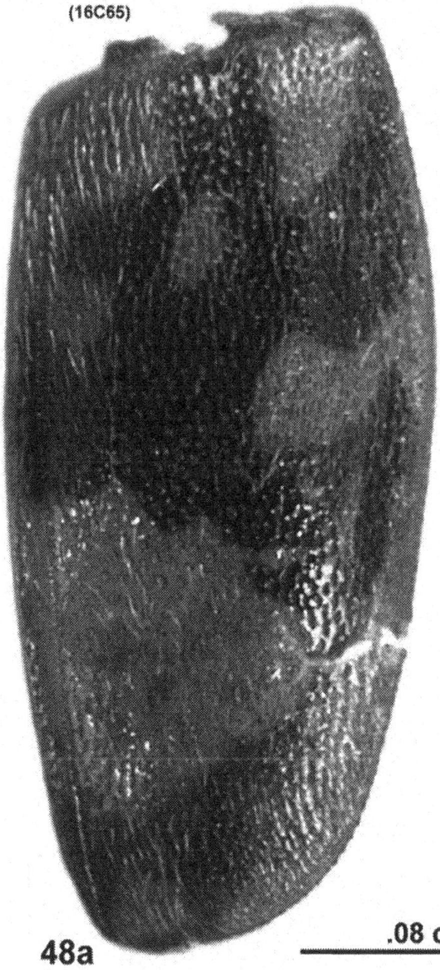

48a

.08 cm

head of *Tenebrio molitor* (Linne)

(16C56)

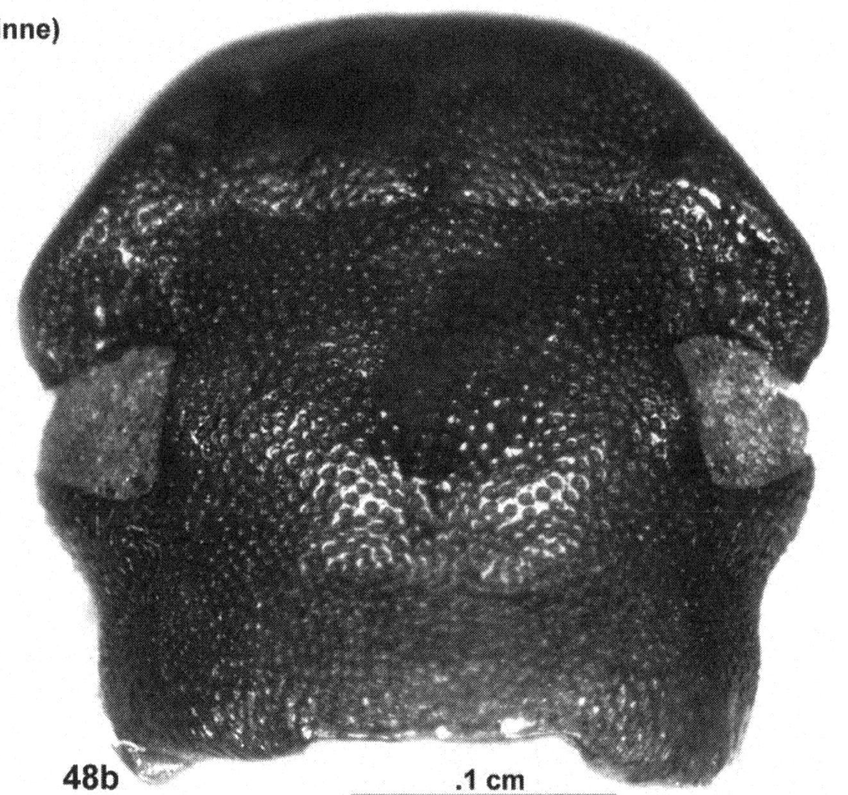

48b

.1 cm

Lytta vesicatoria Linne

(Canadian National Collection)

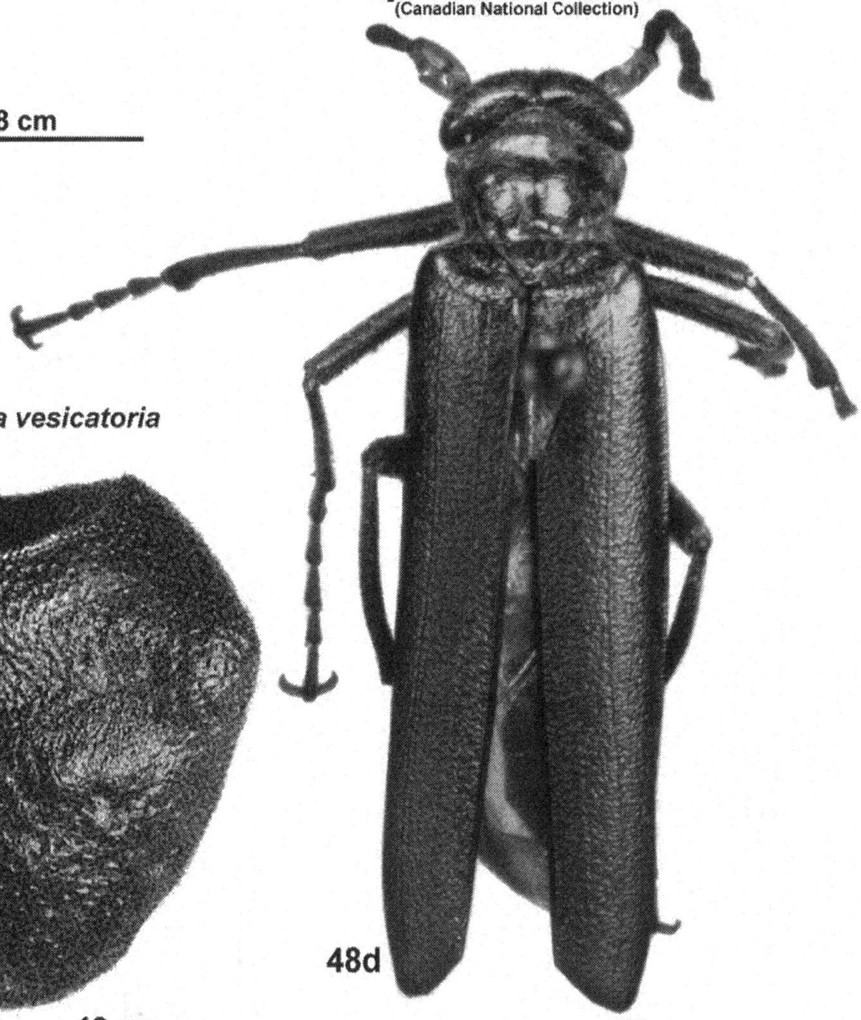

48d

.37 cm

pronotum of cf. *Lytta vesicatoria*

(11C33)

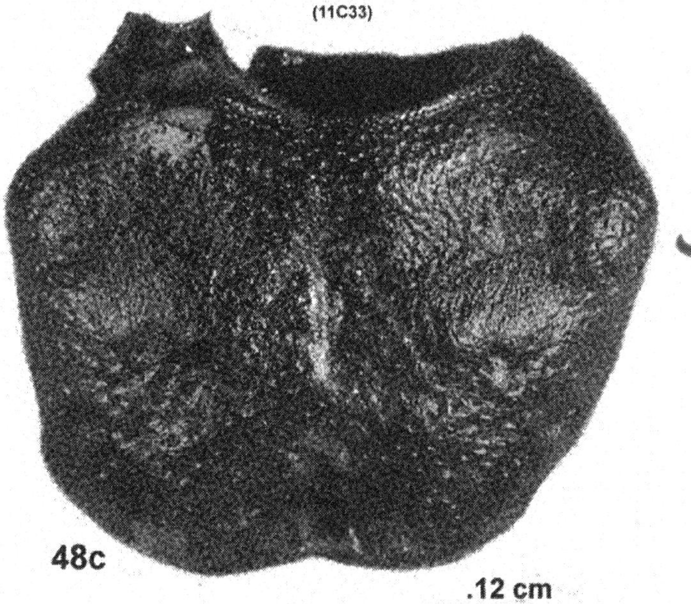

48c

.12 cm

right elytron of *Cercyon unipunctatus* (Linne)

(11C33)

49a .06 cm

head of *Ptinus* sp.

(16C55)

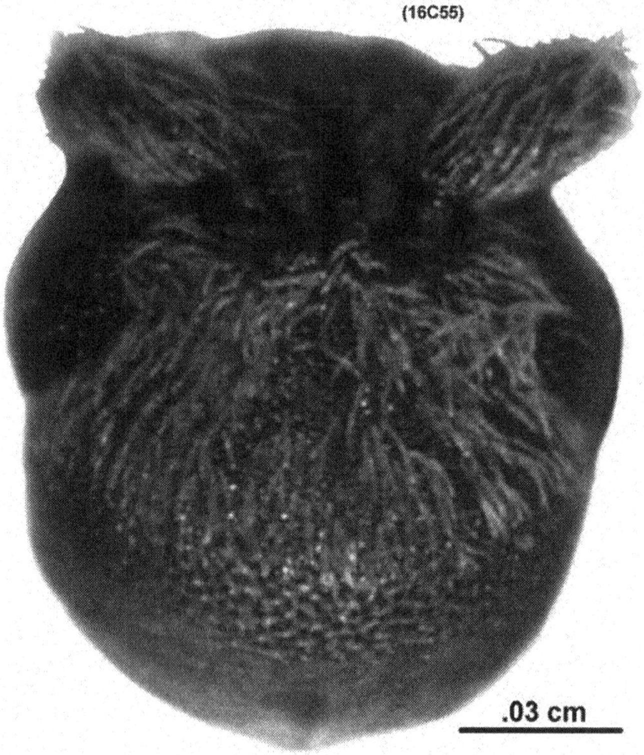

.03 cm

head & pronotum of *Lathridius minutus* Linne group

(16C66)

49c .031 cm

right elytron of *Dermestes lardarius* (Linne)

(16D31)

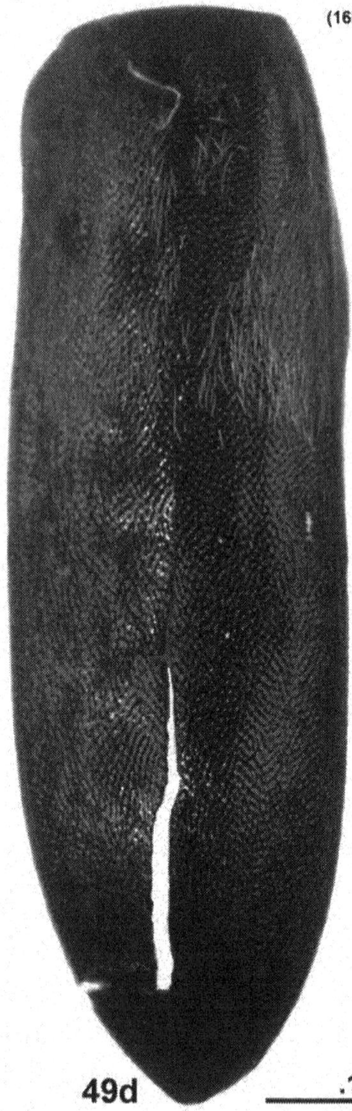

49d .15 cm

APPENDIX B

TABLES

Table 1. Insect remains recovered from privy deposit 16C74 dating to c. 1850

Species	16C74
DICTUOPTERA	
Blatellidae	
Blatella germanica (Linné)+	1
HEMIPTERA	
Cimicidae	
Cimex cf. lectularius Linné+	4
COLEOPTERA	
Carabidae	
Trechus apicalis Motschulsky*	1
B. tetracolum Say+	2
B. versicolor (LeConte)	1
Micropeplidae	
Micropeplus browni Campbell	1
M. tesserula Curtis*	1
Staphylinidae	
Omaliinae spp.	8
Carpelimus obesus (Kiesenwetter)+	1
Oxytelus sculptus Gravenhorst+	13
Anotylus rugosus (Fabricius)+	6
Sepedophilus sp.	1
Aleocharinae spp.	10
Neobisnius villosulus (Stephens)+	1
Neobisnius sp.	2
Philonthus debilis (Gravenhorst)+	1
P. politus (Linné)+	14
P. umbratilis (Gravenhorst)+	3
P. validus Casey	1
P. varians species group	3
Bisnius cephalotes (Gravenhorst)+	2
B. sordidus (Gravenhorst)+	1
Staphylininae spp.	5
Quedius mesomelinus (Marsham)+	9
Quedius sp.	1
Xantholininae spp.	1
Staphylinidae indet. Spp.	9
Hydrophilidae	
C. pygmaeus (Illiger)+	2
Cercyon terminatus (Marsham)+	1
C. unipunctatus (Linné)+	2
C. analis (Paykull)+	28
Cryptopleurum minutum (Fabricius)+	4
Histeridae	
Dendrophilus punctatus (Herbst)+	1
Hydrophilidae or Histeridae larvae	X
Scarabaeidae	
Trox scaber (Linné)+	2
Dermestidae	
Dermestes sp.	1

Species	16C74
Attagenus unicolor japonicus Reitter	1
A. cf. *unicolor japonicus Reitter*	2
Ptinidae	
Ptinus fur (Linné)+	1
Ptinus spp.	6
Ptinidae spp.	9
Nitidulidae	
Nitidulidae indet. spp.	2
Monotomidae	
M. picipes Herbst+	1
M. testacea Motschulsky	1
Cucujidae	
Pediacus sp.	1
Laemophloeus sp.	1
Nausibius clavicornis (Kugelmann)+	1
Oryzaephilus sp.+	2
Cryptophagidae	
Cryptophagus spp.	12
Atomaria sp.	1
Endomychidae	
Mycetaea subterranea (Fabricius)+	2
Lathridiidae	
Lathridius minutus (Linné)+ group	15
Corticaria or *Corticarina* spp.	1
Chrysomelidae	
Gastrophysa polygoni (Linné)+	1
Curculionidae	
Sitophilus oryzae (Linné)+	1
Dryophthorus americanus Bedel	6
Scolytidae	
Polygraphus rufipennis (Kirby)	9
Dryocoetes autographus (Ratzeburg)	1
Scolytidae indet.	1
Coleoptera indet. spp.	32
DIPTERA	
cf. **Ephydridae**	X
Sphaeroceridae	X
Muscoidea	X
Fannia sp.	X
Total	**254**
+ introduced species	
* Holarctic species	
x present	

Table 2. Coleopteran remains recovered from privy 16D25 dating to 1825-1875

Species	16D25
COLEOPTERA	
Carabidae	
Carabidae sp.	1
Staphylinidae	
Oxytelus sculptus Gravenhorst+	1
Anotylus rugosus (Fabricius)+	1
Aleocharinae spp.	8
Staphylinidae indet. spp.	1
Dermestidae	
Attagenus cf. *unicolor japonicus*	1
Ptinidae	
Ptinus spp.	7
Cucujidae	

Species	16D25
Nausibius clavicornis (Kugelmann)+	5
Cucujidae spp.	2
Endomychidae	
Mycetaea subterranea (Fabricius)+	1
Lathridiidae	
Lathridius minutus (Linné)+ group	5
Tenebrionidae	
Tribolium audax Halstead	1
Curculionidae	
Sitophilus sp.	1
Coleoptera indet. spp.	3
Total	**38**
+ introduced species	

Table 3. Insect remains recovered from 16E100 structure dating to c. 1850

Species	16E1	16E2	16E3	16E4	16E5
COLEOPTERA					
Carabidae					
Trechus apicalis					3
Pterostichus corvinus					1
Staphylinidae					
Sepedophilus sp.				1	
Aleocharinae spp.		2	8	4	7
Staphylininae spp.		1	2	3	1
Quedius mesomelinus					1
Xantholininae spp.					1
Staphylinidae indet. spp.	1	1		2	1
Hydrophilidae					
Cercyon sp.			1		
Histeridae					
Dendrophilus punctatus					4
Carcinops pumilio					1
Histeridae spp.				1	2
Hydrophilidae or				X	
Scarabaeidae					
Trox scaber (Linné)+			1	2	1
Ptinidae					
Tipnus unicolor (Piller &					1
Ptinus fur (Linné)+					7

Species	16E1	16E2	16E3	16E4	16E5
Ptinus spp.			1	1	2
Ptinidae spp.	1			2	1
Monotomidae					
Monotoma					3
Monotoma sp.			1		
Cucujidae					
Nausibius clavicornis					1
Cryptophagidae					
Cryptophagus spp.				1	6
Endomychidae					
Mycetaea subterranea				1	8
Lathridiidae					
Lathridius minutus				1	2
Tenebrionidae					
Tenebrio molitor Linné+				1	
Curculionidae					
Cossonus americanus					1
Dryophthorus				7	63
Coleoptera indet. spp.				3	1
Totals	**2**	**4**	**14**	**30**	**119**
+ introduced species					
* Holarctic species					
x present					

Table 4. Insect remains recovered from privy section 16D200 dating to c. 1860

Species	11C29	11C31	11C33	16D27	16D31	16D33	16D35	16D40	16D41
DICTUOPTERA									
Blatellidae									
Blatella germanica (Linné)+					2				
PHTHIRAPTERA									
Phthiraptera indet.					2				
HEMIPTERA									
Cimicidae									
Cimex cf. lectularius Linné+			5	8	8		2		
COLEOPTERA									
Carabidae									
Trechus apicalis Motschulsky*									1
Bembidion scopulinum (Kirby)*			1						

Species	11C29	11C31	11C33	16D27	16D31	16D33	16D35	16D40	16D41
B. tetracolum Say+				5	4				
Bembidion sp.					1			1	
Pterostichus corvinus Dejean			1		1				
Pterostichus spp.			2						
Carabidae sp.				2					
Micropeplidae									
Micropeplus browni Campbell	4	1	36	58	211	1	18	10	16
M. tesserula Curtis*			1	5	2		1		
Silphidae									
Necrodes surinamensis (Fabricius)			1						
Staphylinidae									
Omaliinae spp.				15	17		7	2	6
Hapalaraea floralis (Paykull)+		1	4						
Syntomium grahami Hatch				2					
Carpelimus obesus (Kiesenwetter)+	81	341	10	1					
C. bilineatus Stephens+		7	6	5	1		1		2
Carpelimus sp.				8	4		3		
Oxytelus sculptus Gravenhorst+	28	31	19	13	16		8		1
Anotylus rugosus (Fabricius)+	3	5	4	6	5		3		
Cilea silphoides (Linné)+			7	4	2		2		
Sepedophilus sp.				6	3				1
Aleocharinae spp.				23	30		10	2	1
Neobisnius villosulus (Stephens)+							1		
Neobisnius sp.				1					
Erichsonius pusio (Horn)				2					
Philonthus debilis (Gravenhorst)+				2	2				
P. discoideus (Gravenhorst)+				3	2				
P. politus (Linné)+			3	1	3		1		2
P. umbratilis (Gravenhorst)+				1	5		1		
P. validus Casey			2						
P. varians species group				1	1				
P. ventralis (Gravenhorst)+					1				
Philonthus sp.		1	1		2		3		
Bisnius cephalotes (Gravenhorst)+				2	2		1		
B. sordidus (Gravenhorst)+				1	1		1		
Staphylininae spp.				15	9		8		12
Creophilus maxillosus (Linné)			1						
Quedius mesomelinus (Marsham)+			3	29	12		7		3
Quedius sp.			4	2					
Gyrohypnus angustatus Stephens+				3	3		2	1	2
G. fracticornis (O.F. Müller)+			1	1	1		1		
Gyrohypnus sp.	4	1	3						1
Neohypnus obscurus (Erichson)				2	2				
Xantholininae spp.					1		1		
Staphylinidae indet. spp.				7	8		8	1	4
Hydrophilidae									
Cercyon haemorrhoidalis (Fabricius)+			1		1				
C. pygmaeus (Illiger)+		3	5	3	7		1		
C. terminatus (Marsham)+			1	1	5				
C. unipunctatus (Linné)+			6	1	11		4		1
C. analis (Paykull)+	1	3	21	70	50		17	1	3
Cercyon sp.				18	16	1	7	1	6
Cryptopleurum minutum (Fabricius)+	5		8	1	6		4	1	1
Histeridae									
Margarinotus merdarius (Hoffman)	2	3	5	1			1		1
Margarinotus sp.		1		1					
Atholus bimaculatus (Linné)+	9	17	5			1			
Dendrophilus punctatus (Herbst)+			2		1				1
Carcinops pumilio (Erichson)+				1					
Gnathoncus rotundatus (Kugelmann)+	1		2	1	2				
Histeridae spp.									2
Hydrophilidae or Histeridae larvae		X		X	X		X	X	X
Clambidae									

Species	11C29	11C31	11C33	16D27	16D31	16D33	16D35	16D40	16D41
Clambus armadillo armadillo		1	3						
Scarabaeidae									
Aphodius fimetarius (Linné)+				2	2		1		1
A. granarius (Linné)+				1					
Aphodius sp.		1	1		1				
Trox scaber (Linné)+		1	2	3	5		1		1
Elmidae									
Elmidae			2						
Dermestidae									
Dermestes lardarius Linné+			3		4				
Dermestes sp.				2	1		1		
Attagenus unicolor japonicus Reitter+				5	3				
A. cf. *unicolor japonicus* Reitter+				7	7	1	3		1
Bostrichidae									
Rhyzopertha dominica (Fabricius)+			1						
Anobiidae									
Stegobium paniceum (Linné)+		1	3	30	6		8	1	1
Priobium sericeum (Say)				1			1		1
Ptinidae									
Tipnus unicolor (Piller & Mitterpacher)+				5					
Ptinus fur (Linné)+			2	8	4				
Ptinus spp.				9	18		10	5	4
Ptinidae spp.				13	34	1	8	1	1
Nitidulidae									
Carpophilus hemipterus (Linné)+			1	1	2		2		1
Omosita colon (Linné)+			1						
Omosita sp.					1		1		
Nitidulidae indet. spp.				3	2				
Monotomidae									
Monotoma longicollis (Gyllenhal)+					1				
M. picipes Herbst+			2	1	1				
M. quadrifoveolata Aubé			2		1		1		
M. testacea Motschulsky		1	3	3	5		3		
Monotoma sp.									2
Cucujidae									
Pediacus fuscus Erichson*				3			1		
Pediacus sp.					3				
Laemophloeus bigutattus (Say)				1			2		
Laemophloeus sp.					8			2	
Cryptolestes sp.					6				
Ahasverus advena (Walt)+					1				
Nausibius clavicornis (Kugelmann)+			2	6	6	4	8		2
Oryzaephilus surinamensis (Linné)+				1	1				1
Oryzaephilus sp.+							1		
Cryptophagidae									
Henoticus serratus (Gyllenhal)			1	3	3				
Cryptophagus spp.			8	4	15		4		2
Atomaria sp.				2	3		1		1
Cryptophagidae sp.							1		
Endomychidae									
Mycetaea subterranea (Fabricius)+		1	1	8	13		8		1
Lathridiidae									
Dienerella ruficollis (Marsham)+					1				
D. filiformis (Gyllenhal)+				1	1				
Lathridius minutus (Linné)+ group			1	36	88		29	4	10
cf. *L. minutus* (Linné)+			13						
Cartodere constricta (Gyllenhal)+					1		1		
Corticaria or *Corticarina* spp.	1		1	21	16		1		
Tenebrionidae									
Tribolium castaneum (Herbst)+	1		1						
T. confusum Jacquelin du Val+			1						
Tenebrio molitor Linné+			1	3	2				
T. obscurus Fabricius+			1						

Species	11C29	11C31	11C33	16D27	16D31	16D33	16D35	16D40	16D41
Meloidae									
Lytta cf. *vesicatoria* Linné+			1		1		1		
Salpingidae									
Sphaeriestes virescens (LeConte)			2	1	4		1		
Anthicidae									
Anthicus floralis (Linné)+			1		1				
A. formicarius (Goeze)+			4	1	2		3		
Anthicus sp.			4	2	3				
Chrysomelidae									
Crepidodera sp.			1						
Curculionidae									
Listronotus sp.	1								
Curculio sp.							1		
Sitophilus granarius (Linné)+					1		1		
cf. *Sitophilus granarius* (Linné)+			1						
S. oryzae (Linné)+			2	1	9		3	1	1
Cossonus americanus Buchanan					1				
Dryophthorus americanus Bedel	2		3	2	4		1	1	4
Scolytidae									
Hylurgops rugipennis pinifex (Fitch)							1		
D. simplex LeConte							1		
Polygraphus rufipennis (Kirby)			5		13		6	1	4
Pityokteines sparsus (LeConte)			1		2				
Orthotomicus caelatus (Eichoff)							1		
Xyloterinus politus (Say)					1				
Trypodendron lineatum (Olivier)				1					
Pityophthorus sp.				1					
Scolytidae indet.									1
Coleoptera indet. spp.				67	126		38		10
DIPTERA									
cf. **Ephydridae**				X	X		X		X
Sphaeroceridae					X				
Muscoidea							X		X
Fannia sp.					X				
Totals	**143**	**421**	**253**	**591**	**895**	**9**	**279**	**36**	**117**
+ introduced species									
* Holarctic species									
x present									

Table 5. Insect remains recovered from privy deposit 16C70 dating to c. 1860

Species	16C70
DICTUOPTERA	
Blatellidae	
Blatella germanica (Linné)+	3
PHTHIRAPTERA	
cf. Pediculidae sp.	1
HEMIPTERA	
Nabidae	
Nabidae sp.	1
Cimicidae	
Cimex cf. Lectularius Linné+	3
Anthrocoridae	
Anthrocoridae sp.	1
COLEOPTERA	
Carabidae	
Bembidion tetracolum Say+	1
B. versicolor (LeConte)	1
Syntomus americanus (Dejean)	1
Micropeplidae	
Micropeplus browni Campbell	2

Species	16C70
M. tesserula Curtis*	1
Micropeplus sp.	1
Staphylinidae	
Omaliinae spp.	4
Carpelimus bilineatus Stephens+	5
Carpelimus sp.	2
Oxytelus sculptus Gravenhorst+	10
Cilea silphoides (Linné)+	1
Sepedophilus sp.	2
Aleocharinae spp.	8
Neobisnius villosulus (Stephens)+	1
Neobisnius sp.	1
Philonthus politus (Linné)+	2
P. umbratilis (Gravenhorst)+	4
P. varians species group	1
P. ventralis (Gravenhorst)+	3
Philonthus sp.	2
Staphylininae spp.	4
Quedius mesomelinus (Marsham)+	9

Species	16C70
Gyrohypnus angustatus Stephens+	1
G. fracticornis (O.F. Müller)+	1
Staphylinidae indet. spp.	1
Hydrophilidae	
C. pygmaeus (Illiger)+	1
C. terminatus (Marsham)+	1
C. unipunctatus (Linné)+	4
C. analis (Paykull)+	9
Cercyon sp.	1
Cryptopleurum minutum (Fabricius)+	2
Histeridae	
Gnathoncus rotundatus (Kugelmann)+	1
Hydrophilidae or Histeridae larvae	X
Dermestidae	
Attagenus cf. *unicolor japonicus* Reitter+	3
Ptinidae	
Ptinus fur (Linné)+	3
Ptinus spp.	2
Ptinidae spp.	4
Nitidulidae	
Carpophilus hemipterus (Linné)+	3
Nitidulidae indet. spp.	1
Monotomidae	
Monotoma sp.	2
Cucujidae	
Laemophloeus sp.	1
Nausibius clavicornis (Kugelmann)+	3
Cryptophagidae	
Henoticus serratus (Gyllenhal)	1
Cryptophagus spp.	6
Atomaria sp.	1
Endomychidae	
Mycetaea subterranea (Fabricius)+	4
Lathridiidae	
Lathridius minutus (Linné)+ group	15
Corticaria or *Corticarina* spp.	2
Tenebrionidae	
Tenebrio molitor Linné+	1
Meloidae	
Lytta cf. *vesicatoria* Linné+	1
Anthicidae	
Anthicus floralis (Linné)+	1
A. formicarius (Goeze)+	2
Chrysomelidae	
Phyllotreta striolata (Fabricius)+	1
Curculionidae	
Sitophilus oryzae (Linné)+	2
Dryophthorus americanus Bedel	2
Scolytidae	
Polygraphus rufipennis (Kirby)	2
Orthotomicus caelatus (Eichoff)	1
Coleoptera indet. spp.	26
DIPTERA	
Cf. **Ephydridae**	X
Sphaeroceridae	X
Muscoidea	X
Fannia sp.	X
Total	**187**
+ introduced species	
* Holarctic species	
x present	

Table 6. Coleopteran remains recovered from 16D203 drain dating to c. 1860

Species	16D30	16D38	16K5
COLEOPTERA			
Micropeplidae			
Micropeplus browni Campbell		1	
Staphylinidae			
Omaliinae spp.		2	2
Sepedophilus sp.			1
Aleocharinae spp.		2	1
Philonthus sp.			1
Staphylininae spp.		1	
Quedius mesomelinus (Marsham)+		2	
Gyrohypnus fracticornis (O.F.		1	
Staphylinidae indet. spp.		1	6
Hydrophilidae			
Cercyon analis (Paykull)+		1	1
Cercyon sp.		1	1
Histeridae			
Dendrophilus punctatus (Herbst)+			1
Gnathoncus rotundatus			1
Hydrophilidae or Histeridae		X	X
Scarabaeidae			
Trox scaber (Linné)+		1	1
Ptinidae			
Ptinus spp.	1	2	
Ptinidae spp.			2
Monotomidae			
Monotoma quadrifoveolata Aubé		1	
Cucujidae			
Pediacus sp.		1	
Nausibius clavicornis			1
Cucujidae spp.			1
Curculionidae			
Dryophthorus americanus Bedel		1	1
Scolytidae			
Pityophthorus sp.		1	
Coleoptera indet. spp.	2	3	1
DIPTERA			
cf. **Ephydridae**		X	X
Totals	**3**	**22**	**22**
+ introduced species			
* Holarctic species			
x present			

Table 7. Insect remains from 16C26 privy section dating to c. 1870

Species ⟶	44	47	52	54	55	56	57	59	61	65	66	69	72	75
DICTUOPTERA														
Blatellidae														
Blatella germanica (Linné)+										1	1	1	2	
PHTHIRAPTERA														
cf. Pediculidae sp.													1	
Phthiraptera indet.											2			
HEMIPTERA														
Cimicidae														
Cimex cf. lectularius Linné+									1		2		2	
COLEOPTERA														
Carabidae														
Trechus apicalis Motschulsky*													1	
Bembidion tetracolum Say+			1					1						
B. cf. *tetracolum* Say+			1											
Micropeplidae														
Micropeplus browni Campbell												1		
Micropeplus sp.														1
Silphidae														
Necrodes surinamensis (Fabricius)											1			
Staphylinidae														
Omaliinae spp.	2	4	9	2	5	1	1	1	2	3	8	4	5	2
Carpelimus obesus (Kiesenwetter)+	1	1		64				3						
C. bilineatus Stephens+	12	2	11	28		11	9	67	2	1	1	3		
Carpelimus sp.		5	6	17	2		6			1	1		2	
Oxytelus sculptus Gravenhorst+	21	5	25	19	24	3	7	53	15	14	21	22	9	2
Anotylus rugosus (Fabricius)+	5	1	3	5		1	4	7	1	1	2		2	1
Cilea silphoides (Linné)+		2	5		3	4	1	2			1		2	
Sepedophilus sp.									1		1		1	
Aleocharinae spp.	14	7	41	13	14	8	8	64	10	8	16	3	19	5
Neobisnius villosulus (Stephens)+				1	2	2		1			3		3	
Neobisnius sp.			4					2						
Erichsonius pusio (Horn)						1								
Philonthus debilis (Gravenhorst)+	1										2			
P. discoideus (Gravenhorst)+						1								
P. politus (Linné)+	1	2	4	1		1	1	3	2	2	5	2	4	
P. sericans (Gravenhorst)												1		
P. umbratilis (Gravenhorst)+			1								2			
P. validus Casey		1		2							5			
P. varians species group		1	3	1	1	2		2	1		1			
P. ventralis (Gravenhorst)+					5									
Philonthus sp.	2		7				4	3	1	1	1			
Bisnius cephalotes (Gravenhorst)+					1						3			
B. sordidus (Gravenhorst)+					1						1			
Staphylininae spp.	11	4	7	11	12	5	2	22	12	10	1	3		9
Creophilus maxillosus (Linné)		1					1							
Quedius mesomelinus (Marsham)+	1		2		2	1		1	6	1	11		4	
Quedius sp.				1										
Gyrohypnus angustatus Stephens+	1		1	10	2	1		3		1	1			
G. fracticornis (O.F. Müller)+	2	5	3	2	4	3	2	4	2		1			
Neohypnus obscurus (Erichson)	1	1		1				2			1			
Xantholininae spp.	3	5	5	2				23	4		1			1
Staphylinidae indet. Spp.	1	11	20		13		4	14	2	2	8	2	2	1
Hydrophilidae														
Cercyon haemorrhoidalis			1	1				1				1		
C. pygmaeus (Illiger)+	2	2	11	1	10	8	4	7	6	2	1	2		
C. terminatus (Marsham)+	1	1							1		2			
C. unipunctatus (Linné)+			7		1			2	1	2	6		1	
C. analis (Paykull)+	3	4	8	3	11	2	1	16	13	7	29	1	14	
Cercyon sp.	4	11	14	3	6	2	2	11		4	3		10	2
Cryptopleurum minutum	4	3	4	3	9		3	7	3	1	1	1	1	
Hydrophilidae spp.		1												

Species →	44	47	52	54	55	56	57	59	61	65	66	69	72	75
Histeridae														
Margarinotus merdarius (Hoffman)		1	4	1				1	1					
Margarinotus sp.					1			1			1			
Atholus bimaculatus (Linné)+		5	2	1	1			2				1		
cf. *Atholus* sp.							1					1		
Dendrophilus punctatus (Herbst)+	1					1		1	1		1			
Dendrophilus sp.					1									
Carcinops pumilio (Erichson)+			2											
cf. *Acritus* sp.		1	1		1			5						
Gnathoncus rotundatus		1	3								1			
Histeridae spp.	3	1	3		1								1	
Hydrophilidae or Histeridae	X	X	X	X	X	X	X	X	X	X	X	X	X	
Scarabaeidae														
Aphodius fimetarius (Linné)+			3	1	2			1	1		1	1	1	
Aphodius sp.	2	1		1				1						
Trox scaber (Linné)+	3	3		1	1		1	2	2		2	1		2
Dermestidae														
Dermestes lardarius Linné+											1		1	1
Attagenus unicolor japonicus											1		1	
A. cf. *unicolor japonicus* Reitter+								1	1	1	1		2	
Bostrichidae														
*Rhyzopertha dominica (*Fabricius)+													1	
Anobiidae														
Stegobium paniceum (Linné)+				1				1						
Priobium sericeum (Say)								1						
Ptinidae														
Tipnus unicolor (Piller &			1							1				
Ptinus fur (Linné)+										2	1	1	4	
Ptinus spp.	1	4	12	3	12	1	1	2	3		7	1		5
Ptinidae spp.	2	1	2	3	2	1	2		4		3		7	
Nitidulidae														
Carpophilus hemipterus (Linné)+		1	7		2	1	1	5	5	1	1		1	1
Omosita colon (Linné)+										1	1			
Nitidulidae indet. spp.					1						2	1		
Monotomidae														
Monotoma longicollis (Gyllenhal)+												1		
M. picipes Herbst+			1	1				3	1	1	1	2	3	
M. quadrifoveolata Aubé		1	2	2				1	2	2	2			
M. testacea Motschulsky	3	2	6			1		2	2	1	1		1	
Monotoma sp.	4	2	4	1	4	1	2	5			1	1		
Cucujidae														
Pediacus fuscus Erichson*		1			1						1			
Pediacus sp.				1			2		2					
Laemophloeus sp.					1				1		1		1	
Nausibius clavicornis (Kugelmann)+											2		1	
Cucujidae spp.		1												
Cryptophagidae														
Henoticus serratus (Gyllenhal)			1	1	1						2		1	
Cryptophagus spp.			1	1				2	3	2	12	1	5	
Atomaria sp.									1	1	2			
Cryptophagidae sp.			2											
Endomychidae														
Mycetaea subterranea (Fabricius)+		1	1		2			1	4	4	12	1	5	1
Lathridiidae														
Dienerella ruficollis (Marsham)+													1	
Lathridius minutus (Linné)+ group	3	4	9	2	5	1	1	4	12	3	23	4	11	2
Cartodere constricta (Gyllenhal)+			1											
Corticaria or *Corticarina* spp.			1		2				1		3	1	2	
Tenebrionidae														
cf. *Latheticus oryzae* Waterhouse+									1					
Tenebrio molitor Linné+		1	1	1	1	2			1		2			
Neatus tenebriodes (Palisot de							1							
Salpingidae														

Species ⟶	44	47	52	54	55	56	57	59	61	65	66	69	72	75
Sphaeriestes virescens (LeConte)												1	1	
Anthicidae														
A. formicarius (Goeze)+	1		2				2		3					
Anthicus sp.			1	1	2	3		2	5	2	2	1		
Chrysomelidae														
Hydrothassa vittata (Olivier)*					1									
Curculionidae														
Sitona sp.											1			
Cf. *Hypera* sp.				1										
Sitophilus granarius (Linné)+				1				2	1					
S. oryzae (Linné)+								1			1			
Cossonus americanus Buchanan			1		1									
Dryophthorus americanus Bedel	2	3	4	2	2	1		3	3	3	5		5	
Scolytidae														
Hylurgops rugipennis pinifex (Fitch)											3			
Dendroctinus rufipennis (Kirby)													2	
Polygraphus rufipennis (Kirby)					1				3	1	2		4	1
Dryocoetes affaber (Mannerheim)											2			
Crypturgus borealis Swaine													1	
Xyleborus affinis Eichoff													1	
Scolytidae indet.									2					
Coleoptera indet. spp.	3	10	12	4	13	3		1	23	5	13	7	4	
DIPTERA														
cf. **Ephydridae**	X				X	X		X	X	X	X	X	X	
Sphaeroceridae									X	X	X		X	
Muscoidea	X	X	X	X	X	X	X	X	X	X	X	X	X	
Fannia sp.									X		X			
+ introduced species														
* Holarctic species														
x present														

Table 8. Coleopteran remains recovered from 16D200 privy section dating to c. 1875

Species	16D16	16D17	16D28	16D42
COLEOPTERA				
Micropeplidae				
Micropeplus browni Campbell	1		6	2
Staphylinidae				
Omaliinae spp.	1	4	3	4
Carpelimus obesus (Kiesenwetter)+		1	1	20
C. bilineatus Stephens+				36
Carpelimus sp.		3	2	51
Oxytelus sculptus Gravenhorst+		3	2	45
Anotylus rugosus (Fabricius)+		1	1	8
Cilea silphoides (Linné)+			1	
Aleocharinae spp.		5	4	41
Neobisnius villosulus (Stephens)+				3
Philonthus politus (Linné)+	1			1
P. varians species group				2
Staphylininae spp.		9	3	4
Quedius sp.				1
Gyrohypnus fracticornis (O.F. Müller)+		1	1	2
Gyrohypnus sp.	4			
Xantholininae spp.		2		2
Staphylinidae indet. spp.	1	3		12
Hydrophilidae				
Cercyon pygmaeus (Illiger)+			1	4
C. analis (Paykull)+			2	7
Cercyon sp.		2	3	1
Cryptopleurum minutum (Fabricius)+			1	1

Species	16D16	16D17	16D28	16D42
Histeridae				
Margarinotus merdarius (Hoffman)		1		2
Margarinotus sp.		2	1	1
Atholus bimaculatus (Linné)+			1	20
Gnathoncus rotundatus (Kugelmann)+		1		1
Hydrophilidae or Histeridae larvae	X	X	X	X
Scarabaeidae				
Aphodius fimetarius (Linné)+				1
Trox scaber (Linné)+	2	4	2	1
Anobiidae				
Stegobium paniceum (Linné)+		2	1	1
Priobium sericeum (Say)	1	1	1	2
Ptinidae				
Ptinus spp.		5	4	1
Ptinidae spp.	1	3		1
Nitidulidae				
Carpophilus hemipterus (Linné)+			1	
Nitidulidae indet. spp.			1	
Monotomidae				
Monotoma picipes Herbst+				1
M. quadrifoveolata Aubé		1		
Monotoma sp.			2	
Cucujidae				
Pediacus fuscus Erichson*				1
Pediacus sp.		1		2
Cryptolestes sp.		1	1	
Nausibius clavicornis (Kugelmann)+			1	
Cucujidae spp.		1		
Cryptophagidae				
Henoticus serratus (Gyllenhal)				2
Cryptophagus spp.			1	2
Endomychidae				
Mycetaea subterranea (Fabricius)+			1	
Lathridiidae				
Lathridius minutus (Linné)+ group		1	5	2
Cartodere constricta (Gyllenhal)+		1		
Meloidae				
Lytta cf. *vesicatoria* Linné+		1		
Curculionidae				
Sitophilus granarius (Linné)+				1
Dryophthorus americanus Bedel				1
Scolytidae				
Polygraphus rufipennis (Kirby)			3	
Coleoptera indet. spp.	1	5	2	6
DIPTERA				
cf. **Ephydridae**			X	
Sphaeroceridae			X	
Muscoidea	X			X
Totals	**13**	**65**	**59**	**297**
+ introduced species				
* Holarctic species				
x present				

Table 9. Late nineteenth century insect remains recovered from privy section 16D200

Species	16D18	16D23	16D24	16D26	16D29
INSECTA					
HEMIPTERA					
Cimicidae					
Cimex cf. lectularius Linné+		1		7	
COLEOPTERA					
Carabidae					
Bembidion tetracolum Say+				2	
Poecilus lucublandus (Say)				1	
Micropeplidae					
Micropeplus browni Campbell		4		40	
M. tesserula Curtis*				2	
Micropeplus sp.		2			
Staphylinidae					
Omaliinae spp.	1	4		13	1
Syntomium grahami Hatch	1				
Carpelimus obesus (Kiesenwetter)+				1	
C. bilineatus Stephens+				1	
Carpelimus sp.	1	3		2	
Oxytelus sculptus Gravenhorst+		5		10	
Anotylus rugosus (Fabricius)+	2	6	1	4	1
A. nitidulus (Gravenhorst)+	2				
Sepedophilus sp.				5	
Aleocharinae spp.	1	9		14	2
Neobisnius villosulus (Stephens)+				1	
Philonthus discoideus (Gravenhorst)+				1	
P. politus (Linné)+		1		4	
P. umbratilis (Gravenhorst)+				1	
Bisnius cephalotes (Gravenhorst)+				1	
Staphylininae spp.				5	
Quedius mesomelinus (Marsham)+		1		7	
Gyrohypnus fracticornis (O.F. Müller)+				2	
Neohypnus obscurus (Erichson)				1	
Xantholininae spp.				1	
Staphylinidae indet. spp.	1	2		10	
Hydrophilidae					
Helophorus tuberculatus Gyllenhal*		1			
Cercyon haemorrhoidalis (Fabricius)+		1		1	
C. pygmaeus (Illiger)+				4	
C. unipunctatus (Linné)+				1	
C. analis (Paykull)+	1	6		26	
Cercyon sp.		3	1	5	
Cryptopleurum minutum (Fabricius)+		3	1	1	
Histeridae					
Margarinotus sp.		1			
Atholus bimaculatus (Linné)+				1	
Gnathoncus rotundatus (Kugelmann)+		1		2	
Hydrophilidae or Histeridae larvae	X	X		X	
Scarabaeidae					
Aphodius fimetarius (Linné)+		1			
Trox scaber (Linné)+	2	2		1	
Dermestidae					
Dermestes lardarius Linné+				2	
Attagenus unicolor japonicus Reitter+		7		7	
A. cf. *unicolor japonicus* Reitter+		1			
Anobiidae					
Stegobium paniceum (Linné)+	3	1		8	
Priobium sericeum (Say)	1	1		1	1
Anobiidae sp.			1		
Ptinidae					
Tipnus unicolor (Piller & Mitterpacher)+	1			3	
Niptus hololeucus (Falderman)+		1			
Ptinus fur (Linné)+				1	

Species	16D18	16D23	16D24	16D26	16D29
Ptinus spp.	4	3	1	9	4
Ptinidae spp.				5	
Nitidulidae					
Carpophilus hemipterus (Linné)+		1			
Omosita sp.		1			
Nitidulidae indet. spp.		2		2	
Monotomidae					
Monotoma picipes Herbst+				1	
Monotoma sp.		1		1	
Cucujidae					
Pediacus fuscus Erichson*	4	3	1		
Pediacus sp.					1
Laemophloeus bigutattus (Say)		1		1	
Cryptolestes sp.				1	
Nausibius clavicornis (Kugelmann)+		1		1	
Oryzaephilus sp.+		1		1	
Cucujidae spp.		2			
Cryptophagidae					
Henoticus serratus (Gyllenhal)				6	
Cryptophagus spp.	2			3	
Atomaria sp.				1	
Endomychidae					
Mycetaea subterranea (Fabricius)+	1			4	
Lathridiidae					
Dienerella filiformis (Gyllenhal)+		1			
Lathridius minutus (Linné)+ group	1	4	1	12	
Corticaria or *Corticarina* spp.		2		7	
Tenebrionidae					
Tenebrio molitor Linné+				3	
Salpingidae					
Sphaeriestes virescens (LeConte)				1	
Anthicidae					
Anthicus formicarius (Goeze)+				1	
Bruchidae					
Bruchus sp.+				1	
Curculionidae					
Hypera nigrirostris (Fabricius)+		2		1	
Curculio sp.		1			
Sitophilus granarius (Linné)+		1		1	
Dryophthorus americanus Bedel		1	1	1	
Scolytidae					
Polygraphus rufipennis (Kirby)	2				
Ips sp.				1	
Coleoptera indet. spp.	3	4	1	31	2
DIPTERA					
cf. **Ephydridae**				X	
Totals	**35**	**100**	**9**	**296**	**12**
+ introduced species					
* Holarctic species					
x present					

Table 10. Coleopteran remains recovered from late nineteenth-century renovations to the 1875 building

Species	16C14	16C41
COLEOPTERA		
Staphylinidae		
Staphylinidae indet. spp.	3	
Coleoptera indet. spp.	2	
Total	5	0

Table 11. Coleopteran remains recovered from depression fill below 1875 building

Species	16C16	16C18	16C28
COLEOPTERA			
Carabidae			
B. tetracolum Say +	1		
Carabidae sp.	1		
Micropeplidae			
Micropeplus browni Campbell		1	
M. tesserula Curtis*	1		1
Staphylinidae			
Omaliinae spp.	2	4	1
Carpelimus obesus			2
C. bilineatus Stephens+			4
Carpelimus sp.			2
Oxytelus sculptus			3
Anotylus rugosus (Fabricius)+			3
A. nitidulus (Gravenhorst)+.	1	1	
Aleocharinae spp.	8	1	7
Neobisnius villosulus			1
Philonthus debilis	1		
P. politus (Linné)+			2
Philonthus sp.			1
Staphylininae spp.	1		8
Quedius sp.			1
Gyrohypnus angustatus			1
Xantholininae spp.		1	2
Staphylinidae indet. spp.		3	1
Hydrophilidae			
Helophorus grandis Illiger+		1	
Helophorus sp.		1	
Cercyon haemorrhoidalis			1
C. pygmaeus (Illiger)+	1		2
C. analis (Paykull)+			1
Cercyon sp.	2	1	4
Cryptopleurum minutum			1
Histeridae			
Margarinotus merdarius		1	1
Dendrophilus punctatus			1
Gnathoncus rotundatus	2		1
Histeridae spp.	1		
Hydrophilidae or Histeridae		X	X
Scarabaeidae			
Aphodius fimetarius (Linné)+	1		
Trox scaber (Linné)+	2	3	2
Anobiidae			
Stegobium paniceum (Linné)+		2	
Priobium sericeum (Say)		1	
Anobiidae sp.		1	
Ptinidae			
Ptinus fur (Linné)+	2		
Ptinus spp.	2	2	
Ptinidae spp.		2	1
Nitidulidae			
Omosita colon (Linné)+	1		
Nitidulidae indet. spp.	1		
Monotomidae			
M. picipes Herbst +			1
M. quadrifoveolata Aubé			1
Cucujidae			
Pediacus fuscus Erichson*	2		2
Pediacus sp.		1	
Cucujidae spp.		1	
Lathridiidae			
Lathridius minutus (Linné)+	2	1	
Corticaria or *Corticarina* spp.	2		
Tenebrionidae			
Tenebrio molitor Linné +		1	2
Meloidae			
Lytta cf. *vesicatoria* Linné+		1	
Curculionidae			
Cossonus americanus	2	2	
Dryophthorus americanus	4	2	2
Curculionidae spp.			1
Scolytidae			
Polygraphus rufipennis (Kirby)	2	1	
Coleoptera indet. spp.	6	2	2
DIPTERA			
cf. **Ephydridae**			X
Muscoidea			X
Totals	**51**	**38**	**66**
+ introduced species			
* Holarctic species			
x present			

Table 12. Insect remains recovered from the Abiel Smith School Site, Boston

Species	S1	S2	S3-11	S3-12	S3-13	S3-14	S3-15	S3-16
HEMIPTERA								
Cimicidae								
cf. *Cimex lectularius* Linné +							1	
COLEOPTERA								
Carabidae								
Bembidion cf. *mimus* Hayward						1		
Carabidae sp.							1	
Staphylinidae								
Hapalaraea floralis (Paykull)+		1			1		1	
Arpedium cribratum Fauvel							1	
Carpelimus bilineatus Stephens +		1	5	5	10	21		10
Carpelimus sp.		1						
Oxytelus sculptus Gravenhorst +			1		3	1		2
Anotylus rugosus (Fabricius)+			1		2		1	
Cilea silphoides (Linné)+			2	1	4	1	4	2
Philonthus politus (Linné)+			1				3	
P. discoideus (Gravenhorst)+					1			
P. cf. *discoideus* (Gravenhorst)+						1		
Philonthus sp.			1			1		
Quedius mesomelinus (Marsham)+			1		3		2	2
Quedius sp.					3		2	
Quedius sp. *or Philonthus* sp.							6	1
Gyrohypnus fracticornis (O.F. Müller)+					1			
Staphylinidae indet.	1		7		5	3	4	2
Hydrophilidae								
Cercyon cf. *pygmaeus* (Illiger)+							1	1
C. unipunctatus (Linné)+							1	
Cercyon sp.		1						
Histeridae								
Gnathoncus rotundatus (Kugelmann)+					1		1	
Margarinotus merdarius (Hoffmann)+							1	
Histeridae sp.						1		
Clambidae								
cf. Clambidae sp.			1		1		1	
Scarabaeidae								
Aphodius distinctus (O. F. Müller)+					1			
Ataenius spretulus (Haldeman)			1					
Trox scaber (Linné)+				1				1
Scarabaeidae sp.			1	1		1		
Dermestidae								
Attagenus cf. *unicolor* (Brahm)+		1			1		2	1
Dermestidae sp.							1	
Anobiidae								
Stegobium paniceum (Linné)+			2		5	6	5	1
Ptinidae								
Ptinus fur (Linné)+							3	
cf. *P. fur* (Linné)+				1		1		
Ptinidae sp.			2		2	1		1
Nitidulidae								
cf. *Carpophilus hemipterus* (Linné)+			1					
Stelidota geminata (Say)			1				1	
Nitidulidae sp.	1							
Rhizophagidae								
Monotoma picipes Herbst +						5		1
M. quadrifoveolata Aubé							1	
Cucujidae								
cf. Cryptolestes *sp.*				1			3	1
Oryzaephilus surinamensis *(Linné)*+					1		1	
Oryzaephilus sp.			5		1	2	2	4

Species	S1	S2	S3-11	S3-12	S3-13	S3-14	S3-15	S3-16
Nausibius clavicornis *(Kugelmann)*+			1	2				
cf. *N. clavicornis* (Kugelmann) +					2	1	3	1
cf. Cucujidae sp.					2			
Cryptophagidae								
Cryptophagus sp.			1		2	3		
Endomychidae								
Mycetaea subterranea (Fabricius)+			3		5		8	1
Lathridiidae								
Lathridius minutus group +			1		3	5	6	2
Corticaria sp. *or Corticarina* sp.			1				3	
Tenebrionidae								
Alphitophagus bifasciatus (Say)							1	
Tribolium castaneum (Herbst)+			1	1		2	1	
cf. *T. castaneum* (Herbst)+								1
Anthicidae								
Anthicus sp.		2	1					
cf. Anthicidae sp.							1	
Bruchidae								
Bruchidae sp.					1			
Curculionidae								
Pissodes sp.							1	
Sitophilus granarius (Linné)+						1	1	1
S. oryzae (Linné)+				1	1	1	2	
Hexarthrum ulkei Horn			1	1		2	3	
Scolytidae								
Hylesinus aculeatus Say			1				1	
Pseudopityophthorus minutissimus							1	1
Coleoptera indet.	1	3	4	3	5	2	3	2
Total	**3**	**10**	**48**	**18**	**67**	**63**	**85**	**39**
+ introduced species								

Table 13. Coleopteran remains recovered from the Faubourg Québec Site, Montréal

Species	E116	E166	E187	E193	E197	E201
Carabidae						
Bembidion tetracolum Say+	1					
Pterostichus corvinus (Dejean)			1			
Anisodactylus harrisi LeConte	1					
Sericoda quadripunctata (DeGeer)				1		
cf. *Agonum* sp.	1					
Micropeplidae						
Micropeplus tesserula Curtis	1			1		
Staphylinidae						
Hapalaraea sp.	1					
Carpelimus obesus (Kiesenwetter)+		1	1			
Oxytelus sculptus Gravenhorst+			1	2		
Anotylus rugosus (Fabricius)+	2	1				1
Sepedophilus testaceus (Fabricius) +	2					
Philonthus politus (Linné)+	1					
P. cf. *politus* (Linné)+						1
P. ventralis+	1					
Philonthus sp.	1					
Creophilus maxillosus (Linné)+	1					
Quedius cf *mesomelinus* (Marsham)+	1			1		
Gyrohypnus cf. *fracticornis* (O.F.Müller)+				2		
Gyrohypnus sp.	1					
Staphylinidae spp.		4	3	3	3	1
Hydrophilidae						
Cercyon cf. *haemorrhoidalis* (Fabricius)+				1		
Cercyon cf. *analis* (Paykull)+				4	2	
Cryptopleurum minutum (Fabricius)+				1		

Species	E116	E166	E187	E193	E197	E201
Hydrophilidae sp.	2					
Histeridae						
Gnathoncus nanus (Scriba)+	1					
Clambidae						
Clambus armadillo armadillo (DeGeer)+				5		
Scarabaeidae						
Aphodius cf. *distinctus* (O.F. Müller)+			1	1		
A. granarius (Linné)+			1	2	1	1
Aphodius sp.	1			1		
Trox scaber (Linné)+	1			1	1	
Dermestidae						
cf. *Dermestes lardarius* (Linné)+	1					
Dermestidae spp.			2	1		
Anobiidae						
Stegobium paniceum (Linné) +			1			
Anobiidae spp.	1					
Ptinidae						
Ptinidae sp.	1			1		
Nitidulidae						
Omosita cf. *colon* (Linné)+	1			2		
Nitidulidae sp.				2		
Monotomidae						
Monotoma sp.				1		
Cucujidae						
Nausibius clavicornis (Kugelmann) +			1			
Cryptophagidae						
Cryptophagus spp.	1			3		
Endomychidae						
Mycetaea subterranea (Fabricius) +				1		
Lathridiidae						
Dienerella sp.	1					
cf. *Lathridius minutus* (Linné)+				2		
Corticaria sp. or *Corticarina* sp.	1			3		
Lathridiidae sp.			1			
Tenebrionidae						
Alphitophagus bifasciatus (Say)				1		
Tribolium castaneum (Herbst)+				1		
Anthicidae						
Anthicidae sp.				1		
Chrysomelidae						
Gastrophysa polygoni (Linné)+	1					
Phyllotreta striolata (Fabricius)+	1			1		
cf. *Crepidodera* sp.	1			1		
Chrysomelidae spp.				1		
Curculionidae						
Sitophilus granarius (Linné)+			1			
Dryophthorus americanus Bedel		12	2	4		
Curculionidae sp.	1		1			
Totals	**31**	**18**	**17**	**52**	**7**	**4**
+ introducted speices						

APPENDIX C

COMPLETE LISTING OF ALL INSECT REMAINS RECOVERED FROM THE ÎLOT HUNT SITE (CeEt-110)

Sub-operation	11C			16C													
Level	29	31	33	14	16	18	28	41	44	47	52	54	55	56	57	59	61	65
DICTUOPTERA																		
Blatellidae																		
Blatella germanica (Linné)+																		1
PHTHIRAPTERA																		
cf. Pediculidae sp.																		
Phthiraptera indet.																		
HEMIPTERA																		
Nabidae																		
Nabidae sp.																		
Cimicidae																		
Cimex cf. lectularius Linné+			5														1	
Anthrocoridae																		
Anthrocoridae sp.																		
Total Hemiptera																		
COLEOPTERA																		
Carabidae																		
Trechus apicalis Motschulsky*																		
*Bembidion scopulinum (Kirby)**			1															
B. tetracolum Say +					1						1					1		
B. cf. *tetracolum* Say +											1							
B. versicolor (LeConte)																		
Bembidion sp.																		
Poecilus lucublandus (Say)																		
Pterostichus corvinus Dejean			1															
Pterostichus spp.			2															
Syntomus americanus (Dejean)																		
Carabidae sp.					1													
Total Carabidae																		
Micropeplidae																		
Micropeplus browni Campbell	4	1	36			1												
M. tesserula Curtis*			1		1		1											
Micropeplus sp.																		
Total Micropeplidae																		
Silphidae																		
Necrodes surinamensis (Fabricius)			1															
Staphylinidae																		
Omaliinae spp.					2	4	1		2	4	9	2	5	1	1	1	2	3
Hapalaraea floralis (Paykull)+		1	4															
Syntomium grahami Hatch																		
Carpelimus obesus (Kiesenwetter)+	81	341	10				2		1	1		64				3		
C. bilineatus Stephens+		7	6				4		12	2	11	28		11	9	67	2	1
Carpelimus sp.							2			5	6	17	2		6			1
Oxytelus sculptus Gravenhorst +	28	31	19				3		21	5	25	19	24	3	7	53	15	14
Anotylus rugosus (Fabricius)+	3	5	4				3		5	1	3	5		1	4	7	1	1
A. nitidulus (Gravenhorst)+.					1	1												
Cilea silphoides (Linné)+			7							2	5		3	4	1	2		
Sepedophilus sp.																	1	
Aleocharinae spp.					8	1	7		14	7	41	13	14	8	8	64	10	8
Neobisnius villosulus (Stephens)+							1					1	2	2		1		
Neobisnius sp.											4					2		
Erichsonius pusio (Horn)														1				
Philonthus debilis (Gravenhorst)+					1				1									
P. discoideus (Gravenhorst)+														1				
P. politus (Linné)+			3				2		1	2	4	1		1	1	3	2	2
P.sericans (Gravenhorst)+																		
P. umbratilis (Gravenhorst)+											1							
P. validus Casey			2							1		2						
Philonthus varians species group										1	3	1	1	2		2	1	
P. ventralis (Gravenhorst)+													5					
Philonthus sp.		1	1				1		2		7				4	3	1	1
Bisnius cephalotes (Gravenhorst)+													1					
B. sordidus (Gravenhorst)+													1					
Staphylininae spp.					1		8		11	4	7	11	12	5	2	22	12	10
Creophilus maxillosus (Linné)			1							1					1			
Quedius mesomelinus (Marsham)+			3						1		2		2	1		1	6	1

Sub-operation	11C			16C														
Level	29	31	33	14	16	18	28	41	44	47	52	54	55	56	57	59	61	65
Quedius sp.			4				1					1						
Gyrohypnus angustatus Stephens+							1		1		1	10	2	1		3		1
G. fracticornis (O.F. Müller)+			1						2	5	3	2	4	3	2	4	2	
Gyrohypnus sp.	4	1	3															
Neohypnus obscurus (Erichson)									1	1		1				2		
Xantholininae spp.						1	2		3	5	5	2				23	4	
Staphylinidae indet. spp.				3		3	1		1	11	20		13		4	14	2	2
Total Staphylinidae																		
Hydrophilidae																		
Helophorus grandis Illiger+						1												
H. tuberculatus Gyllenhal*																		
Helophorus sp.						1												
Cercyon haemorrhoidalis (Fabricius)+			1				1				1	1				1		
C. pygmaeus (Illiger)+		3	5		1		2		2	2	11	1	10	8	4	7	6	2
C. terminatus (Marsham)+			1						1	1							1	
C. unipunctatus (Linné)+			6								7		1			2	1	2
C. analis (Paykull)+	1	3	21				1		3	4	8	3	11	2	1	16	13	7
Cercyon sp.					2	1	4		4	11	14	3	6	2	2	11		4
Cryptopleurum minutum (Fabricius)+	5		8				1		4	3	4	3	9		3	7	3	1
Hydrophilidae spp.										1								
Total Hydrophlidae																		
Histeridae																		
Margarinotus merdarius (Hoffman)	2	3	5			1	1			1	4	1				1	1	
Margarinotus sp.		1											1			1		
Atholus bimaculatus (Linné)+	9	17	5							5	2	1	1			2		
cf. *Atholus* sp.															1			
Dendrophilus punctatus (Herbst)+			2				1		1					1		1	1	
Dendrophilus sp.													1					
Carcinops pumilio (Erichson)+											2							
cf. *Acritus* sp.										1	1		1			5		
Gnathoncus rotundatus (Kugelmann)+	1		2		2		1			1	3							
Histeridae spp.					1				3	1	3		1					
Total Histeridae																		
Hydrophilidae or Histeridae larvae		x				x	x		x	x	x	x	x	x	x	x	x	x
Clambidae																		
Clambus armadillo armadillo (DeGeer)+		1	3															
Scarabaeidae																		
Aphodius fimetarius (Linné)+					1						3	1	2			1	1	
A. granarius (Linné)+																		
Aphodius sp.		1	1						2	1		1				1		
Trox scaber (Linné)+		1	2		2	3	2		3	3		1	1		1	2	2	
Total Scarabaeidae																		
Elmidae																		
Optioservus sp.			1															
Elmidae indet. sp.			1															
Dermestidae																		
Dermestes lardarius Linné+			3															
Dermestes sp.																		
Attagenus unicolor janponicus Reitter+																		
A. cf. *unicolor janponicus Reitter*+																1	1	1
Total Dermestidae																		
Bostrichidae																		
Rhyzopertha dominica (Fabricius)+			1															
Anobiidae																		
Stegobium paniceum (Linné)+		1	3			2						1				1		
Priobium sericeum (Say)						1										1		
Anobiidae sp.						1												
Total Anobiidae																		
Ptinidae																		
Tipnus unicolor (Piller & Mitterpacher)+											1							1
Niptus hololeucus (Falderman)+																		
Ptinus fur (Linné)+			2		2													2
Ptinus spp.					2	2			1	4	12	3	12	1	1	2	3	
Ptinidae spp.						2	1		2	1	2	3	2	1	2		4	
Total Ptinidae																		
Nitidulidae																		
Carpophilus hemipterus (Linné)+			1							1	7		2	1	1	5	5	1

Sub-operation	11C			16C														
Level	**29**	**31**	**33**	**14**	**16**	**18**	**28**	**41**	**44**	**47**	**52**	**54**	**55**	**56**	**57**	**59**	**61**	**65**
Omosita colon (Linné)+			1		1													1
Omosita sp.																		
Nitidulidae indet. spp.					1								1					
Total Nitidulidae																		
Monotomidae																		
Monotoma longicollis (Gyllenhal)+																		
M. picipes Herbst +			2				1				1	1				3	1	1
M. quadrifoveolata Aubé			2				1			1	2	2				1	2	2
M. testacea Motschulsky		1	3						3	2	6			1		2	2	1
Monotoma sp.									4	2	4	1	4	1	2	5		
Total Monotomidae																		
Cucujidae																		
Pediacus fuscus Erichson*					2		2			1			1					
Pediacus sp.						1						1			2		2	
Laemophloeus bigutattus (Say)																		
Laemophloeus sp.													1				1	
Cryptolestes sp.																		
Ahasverus advena (Walt)+																		
Nausibius clavicornis (Kugelmann) +			2															
Oryzaephilus suranamensis (Linné)+																		
Oryzaephilus sp.+																		
Cucujidae spp.						1				1								
Total Cucujidae																		
Cryptophagidae																		
Henoticus serratus (Gyllenhal)			1								1	1	1					
Cryptophagus spp.			8								1	1				2	3	2
Atomaria sp.																	1	1
Cryptophagidae sp.											2							
Total Cryptophagidae																		
Endomychidae																		
Mycetaea subterranea (Fabricius)+		1	1							1	1		2			1	4	4
Lathridiidae																		
Dienerella filiformis (Gyllenhal)+																		
D. ruficollis (Marsham)+																		
Lathridius minutus (Linné)+ group			1		2	1			3	4	9	2	5	1	1	4	12	3
cf. *L. minutus* (Linné)+			13															
Cartodere constricta (Gyllenhal)+											1							
Corticaria or *Corticarina* spp.	1		1		2						1		2				1	
Total Lathridiidae																		
Tenebrionidae																		
Tribolium audax Halstead																		
T. castaneum (Herbst)+	1		1															
T. confusum Jacquelin du Val +			1															
cf. *Latheticus oryzae* Waterhouse+																	1	
Tenebrio molitor Linné +			1			1	2			1	1	1	1	2			1	
T. obscurus Fabricius +			1															
Neatus tenebriodes (Palisot de Beauvois)															1			
Total Tenebrionidae																		
Meloidae																		
Lytta cf. *vesicatoria* Linné+			1			1												
Salpingidae																		
Sphaeriestes virescens (LeConte)			2															
Anthicidae																		
Anthicus floralis (Linné)+			1															
A. formicarius (Goeze)+			4						1		2				2		3	
Anthicus sp.			4								1	1	2	3		2	5	2
Total Anthicidae																		
Bruchidae																		
Bruchus sp.+																		
Chrysomelidae																		
Gastrophysa polygoni (Linné)+																		
Hydrothassa vittata (Olivier)*													1					
Phyllotreta striolata (Fabricius)+																		
Crepidodera sp.			1															
Total Chrysomelidae																		
Curculionidae																		
Listronotus sp.	1																	

Sub-operation	11C			16C														
Level	29	31	33	14	16	18	28	41	44	47	52	54	55	56	57	59	61	65
Sitona sp.																		
Hypera nigrirostris (Fabricius)+																		
cf. *Hypera* sp.												1						
Curculio sp.																		
Sitophilus granarius (Linné)+												1				2	1	
cf. *Sitophilus granarius* (Linné)+			1															
S. oryzae (Linné)+			2													1		
Sitophilus sp.																		
Cossonus americanus Buchanan					2	2					1		1					
Dryophthorus americanus Bedel	2		3		4	2	2		2	3	4	2	2	1		3	3	3
Curculionidae indet. sp.							1											
Total Curculionidae																		
Scolytidae																		
Hylurgops rugipennis pinifex (Fitch)																		
Dendroctinus rufipennis (Kirby)																		
D. simplex LeConte																		
Polygraphus rufipennis (Kirby)			5		2	1							1				3	1
Pityokteines sparsus (LeConte)			1															
Orthotomicus caelatus (Eichoff)																		
Ips sp.																		
Dryocoetes affaber (Mannerheim)																		
D. autographus (Ratzeburg)																		
Crypturgus borealis Swaine																		
Xyloterinus politus (Say)																		
Trypodendron lineatum (Olivier)																		
Xyleborus affinis Eichoff																		
Pityophthorus sp.																		
Scolytidae indet.																	2	
Total Scolytidae																		
Coleoptera indet. spp.				2	6	2	2		3	10	12	4	13	3		1	23	5
DIPTERA																		
cf. **Ephydridae**							x		x				x	x		x	x	x
Sphaeroceridae																	x	x
Muscoidea							x		x	x	x	x	x	x	x	x	x	x
Muscidae																		
Fannia sp.																	x	
Total/level	**143**	**421**	**253**	**5**	**51**	**38**	**66**	**0**	**121**	**125**	**294**	**222**	**190**	**73**	**74**	**373**	**175**	**93**

\+ introduced species

* Holarctic species

x present

Sub-operation	16C						16D											
Level	66	69	70	72	74	75	16	17	18	23	24	25	26	27	28	29	30	31
DICTUOPTERA																		
Blatellidae																		
Blatella germanica (Linné)+	1	1	3	2	1													2
PHTHIRAPTERA																		
cf. Pediculidae sp.			1	1														
Phthiraptera indet.	2																	2
HEMIPTERA																		
Nabidae																		
Nabidae sp.			1															
Cimicidae																		
Cimex cf. lectularius Linné+	2		3	2	4					1			7	8				8
Anthrocoridae																		
Anthrocoridae sp.			1															
Total Hemiptera																		
COLEOPTERA																		
Carabidae																		
Trechus apicalis Motschulsky*				1	1													
*Bembidion scopulinum (Kirby)**																		
B. tetracolum Say +			1		2								2	5				4
B. cf. *tetracolum* Say +																		
B. versicolor (LeConte)			1		1													
Bembidion sp.																		1
Poecilus lucublandus (Say)													1					
Pterostichus corvinus Dejean																		1
Pterostichus spp.																		
Syntomus americanus (Dejean)			1															
Carabidae sp.									1			1		2				
Total Carabidae																		
Micropeplidae																		
Micropeplus browni Campbell		1	2		1		1			4			40	58	6			211
M. tesserula Curtis*			1		1								2	5				2
Micropeplus sp.			1			1				2								
Total Micropeplidae																		
Silphidae																		
Necrodes surinamensis (Fabricius)	1																	
Staphylinidae																		
Omaliinae spp.	8	4	4	5	8	2	1	4	1	4			13	15	3	1		17
Hapalaraea floralis (Paykull)+																		
Syntomium grahami Hatch									1					2				
Carpelimus obesus (Kiesenwetter)+					1			1					1	1	1			
C. bilineatus Stephens+	1	3	5										1	5				1
Carpelimus sp.	1		2	2				3	1	3			2	8	2			4
Oxytelus sculptus Gravenhorst +	21	22	10	9	13	2		3		5		1	10	13	2			16
Anotylus rugosus (Fabricius)+	2			2	6	1		1	2	6	1	1	4	6	1	1		5
A. nitidulus (Gravenhorst)+.									2									
Cilea silphoides (Linné)+	1		1	2										4	1			2
Sepedophilus sp.	1		2	1	1								5	6				3
Aleocharinae spp.	16	3	8	19	10	5		5	1	9		8	14	23	4	2		30
Neobisnius villosulus (Stephens)+	3		1	3	1								1					
Neobisnius sp.			1		2									1				
Erichsonius pusio (Horn)														2				
Philonthus debilis (Gravenhorst)+	2				1									2				2
P. discoideus (Gravenhorst)+													1	3				2
P. politus (Linné)+	5	2	2	4	14		1			1			4	1				3
P.sericans (Gravenhorst)+		1																
P. umbratilis (Gravenhorst)+	2		4		3								1	1				5
P. validus Casey	5				1													
Philonthus varians species group	1		1		3									1				1
P. ventralis (Gravenhorst)+			3															1
Philonthus sp.	1		2															2
Bisnius cephalotes (Gravenhorst)+	3				2								1	2				2
B. sordidus (Gravenhorst)+	1				1									1				1
Staphylininae spp.	1	3	4		5	9		9					5	15	3			9
Creophilus maxillosus (Linné)																		
Quedius mesomelinus (Marsham)+	11		9	4	9					1			7	29				12

Sub-operation	16C						16D											
Level	**66**	**69**	**70**	**72**	**74**	**75**	**16**	**17**	**18**	**23**	**24**	**25**	**26**	**27**	**28**	**29**	**30**	**31**
Quedius sp.					1									2				
Gyrohypnus angustatus Stephens+	1		1											3				3
G. fracticornis (O.F. Müller)+	1		1					1					2	1	1			1
Gyrohypnus sp.							4											
Neohypnus obscurus (Erichson)	1												1	2				2
Xantholininae spp.	1				1	1		2					1					1
Staphylinidae indet. spp.	8	2	1	2	9	1	1	3	1	2		1	10	7				8
Total Staphylinidae																		
Hydrophilidae																		
Helophorus grandis Illiger+																		
H. tuberculatus Gyllenhal*										1								
Helophorus sp.																		
Cercyon haemorrhoidalis (Fabricius)+		1								1			1					1
C. pygmaeus (Illiger)+	1	2	1		2								4	3	1			7
C. terminatus (Marsham)+	2		1		1									1				5
C. unipunctatus (Linné)+	6		4	1	2								1	1				11
C. analis (Paykull)+	29	1	9	14	28				1	6			26	70	2			50
Cercyon sp.	3		1	10		2		2		3	1		5	18	3			16
Cryptopleurum minutum (Fabricius)+	1	1	2	1	4					3	1		1	1	1			6
Hydrophilidae spp.																		
Total Hydrophlidae																		
Histeridae																		
Margarinotus merdarius (Hoffman)								1						1				
Margarinotus sp.	1							2		1				1	1			
Atholus bimaculatus (Linné)+		1											1		1			
cf. *Atholus* sp.		1																
Dendrophilus punctatus (Herbst)+	1				1													1
Dendrophilus sp.																		
Carcinops pumilio (Erichson)+														1				
cf. *Acritus* sp.																		
Gnathoncus rotundatus (Kugelmann)+	1		1					1		1			2	1				2
Histeridae spp.				1														
Total Histeridae																		
Hydrophilidae or Histeridae larvae	x	x	x	x	x		x	x	x	x			x	x	x			x
Clambidae																		
Clambus armadillo armadillo (DeGeer)+																		
Scarabaeidae																		
Aphodius fimetarius (Linné)+	1	1		1						1				2				2
A. granarius (Linné)+														1				
Aphodius sp.																		1
Trox scaber (Linné)+	2	1			2	2	2	4	2	2			1	3	2			5
Total Scarabaeidae																		
Elmidae																		
Optioservus sp.																		
Elmidae indet. sp.																		
Dermestidae																		
Dermestes lardarius Linné+	1			1		1							2					4
Dermestes sp.					1									2				1
Attagenus unicolor janponicus Reitter+	1			1	1					7			7	5				3
A. cf. *unicolor janponicus Reitter*+	1		3	2	2					1		1		7				7
Total Dermestidae																		
Bostrichidae																		
Rhyzopertha dominica (Fabricius)+				1														
Anobiidae																		
Stegobium paniceum (Linné)+								2	3	1			8	30	1			6
Priobium sericeum (Say)							1	1	1	1			1	1	1	1		
Anobiidae sp.											1							
Total Anobiidae																		
Ptinidae																		
Tipnus unicolor (Piller & Mitterpacher)+									1				3	5				
Niptus hololeucus (Falderman)+										1								
Ptinus fur (Linné)+	1	1	3	4	1								1	8				4
Ptinus spp.	7	1	2		6	5		5	4	3	1	7	9	9	4	4	1	18
Ptinidae spp.	3		4	7	9		1	3					5	13				34
Total Ptinidae																		
Nitidulidae																		
Carpophilus hemipterus (Linné)+	1		3	1		1				1				1	1			2

Sub-operation	16C						16D											
Level	66	69	70	72	74	75	16	17	18	23	24	25	26	27	28	29	30	31
Omosita colon (Linné)+	1																	
Omosita sp.										1								1
Nitidulidae indet. spp.	2	1	1		2					2			2	3	1			2
Total Nitidulidae																		
Monotomidae																		
Monotoma longicollis (Gyllenhal)+		1																1
M. picipes Herbst +	1	2		3	1								1	1				1
M. quadrifoveolata Aubé	2							1										1
M. testacea Motschulsky	1			1	1									3				5
Monotoma sp.	1	1	2							1			1		2			
Total Monotomidae																		
Cucujidae																		
Pediacus fuscus Erichson*	1								4	3	1			3				
Pediacus sp.					1			1								1		3
Laemophloeus bigutattus (Say)										1			1	1				
Laemophloeus sp.	1		1	1	1													8
Cryptolestes sp.								1					1		1			6
Ahasverus advena (Walt)+																		1
Nausibius clavicornis (Kugelmann) +	2		3	1	1					1		5	1	6	1			6
Oryzaephilus suranamensis (Linné)+														1				1
Oryzaephilus sp.+					2					1			1					
Cucujidae spp.								1		2		2						
Total Cucujidae																		
Cryptophagidae																		
Henoticus serratus (Gyllenhal)	2		1	1									6	3				3
Cryptophagus spp.	12	1	6	5	12				2				3	4	1			15
Atomaria sp.	2		1		1								1	2				3
Cryptophagidae sp.																		
Total Cryptophagidae																		
Endomychidae																		
Mycetaea subterranea (Fabricius)+	12	1	4	5	2	1			1			1	4	8	1			13
Lathridiidae																		
Dienerella filiformis (Gyllenhal)+										1				1				1
D. ruficollis (Marsham)+				1														1
Lathridius minutus (Linné)+ group	23	4	15	11	15	2		1	1	4	1	5	12	36	5			88
cf. *L. minutus* (Linné)+																		
Cartodere constricta (Gyllenhal)+								1										1
Corticaria or *Corticarina* spp.	3	1	2	2	1					2			7	21				16
Total Lathridiidae																		
Tenebrionidae																		
Tribolium audax Halstead												1						
T. castaneum (Herbst)+																		
T. confusum Jacquelin du Val +																		
cf. *Latheticus oryzae* Waterhouse+																		
Tenebrio molitor Linné +	2		1										3	3				2
T. obscurus Fabricius +																		
Neatus tenebriodes (Palisot de Beauvois)																		
Total Tenebrionidae																		
Meloidae																		
Lytta cf. *vesicatoria* Linné+			1					1										1
Salpingidae																		
Sphaeriestes virescens (LeConte)		1		1									1	1				4
Anthicidae																		
Anthicus floralis (Linné)+			1															1
A. formicarius (Goeze)+			2										1	1				2
Anthicus sp.	2	1												2				3
Total Anthicidae																		
Bruchidae																		
Bruchus sp.+													1					
Chrysomelidae																		
Gastrophysa polygoni (Linné)+					1													
Hydrothassa vittata (Olivier)*																		
Phyllotreta striolata (Fabricius)+			1															
Crepidodera sp.																		
Total Chrysomelidae																		
Curculionidae																		
Listronotus sp.																		

Sub-operation	16C						16D											
Level	**66**	**69**	**70**	**72**	**74**	**75**	**16**	**17**	**18**	**23**	**24**	**25**	**26**	**27**	**28**	**29**	**30**	**31**
Sitona sp.	1																	
Hypera nigrirostris (Fabricius)+										2			1					
cf. *Hypera* sp.																		
Curculio sp.										1								
Sitophilus granarius (Linné)+										1			1					1
cf. *Sitophilus granarius* (Linné)+																		
S. oryzae (Linné)+	1		2		1									1				9
Sitophilus sp.												1						
Cossonus americanus Buchanan																		1
Dryophthorus americanus Bedel	5		2	5	6					1	1		1	2				4
Curculionidae indet. sp.																		
Total Curculionidae																		
Scolytidae																		
Hylurgops rugipennis pinifex (Fitch)	3																	
Dendroctinus rufipennis (Kirby)				2														
D. simplex LeConte																		
Polygraphus rufipennis (Kirby)	2		2	4	9	1			2						3			13
Pityokteines sparsus (LeConte)																		2
Orthotomicus caelatus (Eichoff)			1															
Ips sp.													1					
Dryocoetes affaber (Mannerheim)	2																	
D. autographus (Ratzeburg)					1													
Crypturgus borealis Swaine				1														
Xyloterinus politus (Say)																		1
Trypodendron lineatum (Olivier)														1				
Xyleborus affinis Eichoff				1														
Pityophthorus sp.														1				
Scolytidae indet.					1													
Total Scolytidae																		
Coleoptera indet. spp.	13	7	26	4	32		1	5	3	4	1	3	31	67	2	2	2	126
DIPTERA																		
cf. **Ephydridae**	x	x	x	x	x								x	x	x			x
Sphaeroceridae	x		x	x	x										x			x
Muscoidea	x	x	x	x	x		x											
Muscidae																		
Fannia sp.	x		x		x													x
Total/level	**264**	**74**	**187**	**153**	**254**	**37**	**13**	**65**	**35**	**100**	**9**	**38**	**296**	**591**	**59**	**12**	**3**	**897**

\+ introduced species

* Holarctic species

x present

Sub-operation	16D						16E					16K	Total/taxon
Level	33	35	38	40	41	42	1	2	3	4	5	5	
DICTUOPTERA													
Blatellidae													
Blatella germanica (Linné)+													**11**
PHTHIRAPTERA													
cf. Pediculidae sp.													**2**
Phthiraptera indet.													**4**
HEMIPTERA													
Nabidae													
Nabidae sp.													**1**
Cimicidae													
Cimex cf. lectularius Linné+		2											**43**
Anthrocoridae													
Anthrocoridae sp.													**1**
Total Hemiptera													**45**
COLEOPTERA													
Carabidae													
Trechus apicalis Motschulsky*					1						3		**6**
*Bembidion scopulinum (Kirby)**													**1**
B. tetracolum Say +													**17**
B. cf. *tetracolum* Say +													**1**
B. versicolor (LeConte)													**2**
Bembidion sp.				1									**2**
Poecilus lucublandus (Say)													**1**
Pterostichus corvinus Dejean											1		**3**
Pterostichus spp.													**2**
Syntomus americanus (Dejean)													**1**
Carabidae sp.													**5**
Total Carabidae													**41**
Micropeplidae													
Micropeplus browni Campbell	1	18	1	10	16	2							**414**
M. tesserula Curtis*		1											**15**
Micropeplus sp.													**4**
Total Micropeplidae													**433**
Silphidae													
Necrodes surinamensis (Fabricius)													**2**
Staphylinidae													
Omaliinae spp.		7	2	2	6	4						2	**150**
Hapalaraea floralis (Paykull)+													**5**
Syntomium grahami Hatch													**3**
Carpelimus obesus (Kiesenwetter)+						20							**528**
C. bilineatus Stephens+		1			2	36							**215**
Carpelimus sp.		3				51							**121**
Oxytelus sculptus Gravenhorst +		8			1	45							**448**
Anotylus rugosus (Fabricius)+		3				8							**93**
A. nitidulus (Gravenhorst)+.													**4**
Cilea silphoides (Linné)+		2											**37**
Sepedophilus sp.					1					1		1	**23**
Aleocharinae spp.		10	2	2	1	41		2	8	4	7	1	**438**
Neobisnius villosulus (Stephens)+		1				3							**20**
Neobisnius sp.													**10**
Erichsonius pusio (Horn)													**3**
Philonthus debilis (Gravenhorst)+													**9**
P. discoideus (Gravenhorst)+													**7**
P. politus (Linné)+		1			2	1							**63**
P.sericans (Gravenhorst)+													**1**
P. umbratilis (Gravenhorst)+		1											**18**
P. validus Casey													**11**
Philonthus varians species group						2							**20**
P. ventralis (Gravenhorst)+													**9**
Philonthus sp.		3										1	**30**
Bisnius cephalotes (Gravenhorst)+		1											**12**
B. sordidus (Gravenhorst)+		1											**6**
Staphylininae spp.		8	1		12	4		1	2	3	1		**200**
Creophilus maxillosus (Linné)													**3**
Quedius mesomelinus (Marsham)+		7	2		3						1		**112**

Sub-operation	16D						16E					16K	Total/taxon
Level	**33**	**35**	**38**	**40**	**41**	**42**	**1**	**2**	**3**	**4**	**5**	**5**	
Quedius sp.						1							**10**
Gyrohypnus angustatus Stephens+		2		1	2								**33**
G. fracticornis (O.F. Müller)+		1	1			2							**40**
Gyrohypnus sp.					1								**13**
Neohypnus obscurus (Erichson)													**11**
Xantholininae spp.		1				2					1		**56**
Staphylinidae indet. spp.		8	1	1	4	12	1	1		2	1	6	**167**
Total Staphylinidae													**2929**
Hydrophilidae													
Helophorus grandis Illiger+													**1**
H. tuberculatus Gyllenhal*													**1**
Helophorus sp.													**1**
Cercyon haemorrhoidalis (Fabricius)+													**9**
C. pygmaeus (Illiger)+		1				4							**90**
C. terminatus (Marsham)+													**14**
C. unipunctatus (Linné)+		4			1								**50**
C. analis (Paykull)+		17	1	1	3	7						1	**360**
Cercyon sp.	1	7	1	1	6	1			1			1	**147**
Cryptopleurum minutum (Fabricius)+		4		1	1	1							**80**
Hydrophilidae spp.													**1**
Total Hydrophlidae													**754**
Histeridae													
Margarinotus merdarius (Hoffman)		1			1	2							**26**
Margarinotus sp.						1							**10**
Atholus bimaculatus (Linné)+	1					20							**66**
cf. *Atholus* sp.													**2**
Dendrophilus punctatus (Herbst)+					1						4	1	**16**
Dendrophilus sp.													**1**
Carcinops pumilio (Erichson)+											1		**4**
cf. *Acritus* sp.													**8**
Gnathoncus rotundatus (Kugelmann)+						1						1	**21**
Histeridae spp.					2	1				1	2		**16**
Total Histeridae													**170**
Hydrophilidae or Histeridae larvae		x	x	x	x	x				x		x	
Clambidae													
Clambus armadillo armadillo (DeGeer)+													**4**
Scarabaeidae													
Aphodius fimetarius (Linné)+		1			1	1							**20**
A. granarius (Linné)+													**1**
Aphodius sp.													**8**
Trox scaber (Linné)+		1	1		1	1			1	2	1	1	**60**
Total Scarabaeidae													**89**
Elmidae													
Optioservus sp.													**1**
Elmidae indet. sp.													**1**
Dermestidae													
Dermestes lardarius Linné+													**12**
Dermestes sp.		1											**5**
Attagenus unicolor janponicus Reitter+													**25**
A. cf. *unicolor janponicus Reitter*+	1	3			1								**32**
Total Dermestidae													**74**
Bostrichidae													
Rhyzopertha dominica (Fabricius)+													**2**
Anobiidae													
Stegobium paniceum (Linné)+		8		1	1	1							**70**
Priobium sericeum (Say)		1			1	2							**14**
Anobiidae sp.													**2**
Total Anobiidae													**86**
Ptinidae													
Tipnus unicolor (Piller & Mitterpacher)+											1		**12**
Niptus hololeucus (Falderman)+													**1**
Ptinus fur (Linné)+											7		**36**
Ptinus spp.		10	2	5	4	1			1	1	2		**155**
Ptinidae spp.	1	8		1	1	1	1			2	1	2	**117**
Total Ptinidae													**321**
Nitidulidae													
Carpophilus hemipterus (Linné)+		2			1								**38**

Sub-operation	16D						16E					16K	Total/taxon
Level	33	35	38	40	41	42	1	2	3	4	5	5	
Omosita colon (Linné)+													**4**
Omosita sp.		1											**3**
Nitidulidae indet. spp.													**18**
Total Nitidulidae													**63**
Monotomidae													
Monotoma longicollis (Gyllenhal)+													**2**
M. picipes Herbst +						1							**21**
M. quadrifoveolata Aubé		1	1								3		**22**
M. testacea Motschulsky		3											**35**
Monotoma sp.					2				1				**34**
Total Monotomidae													**114**
Cucujidae													
Pediacus fuscus Erichson*		1				1							**20**
Pediacus sp.			1			2							**15**
Laemophloeus bigutattus (Say)		2											**5**
Laemophloeus sp.				2									**16**
Cryptolestes sp.													**9**
Ahasverus advena (Walt)+													**1**
Nausibius clavicornis (Kugelmann) +	4	8			2						1	1	**45**
Oryzaephilus suranamensis (Linné)+					1								**3**
Oryzaephilus sp.+		1											**5**
Cucujidae spp.												1	**8**
Total Cucujidae													**127**
Cryptophagidae													
Henoticus serratus (Gyllenhal)						2							**22**
Cryptophagus spp.		4			2	2				1	6		**93**
Atomaria sp.		1			1								**14**
Cryptophagidae sp.		1											**3**
Total Cryptophagidae													**132**
Endomychidae													
Mycetaea subterranea (Fabricius)+		8			1					1	8		**86**
Lathridiidae													
Dienerella filiformis (Gyllenhal)+													**3**
D. ruficollis (Marsham)+													**2**
Lathridius minutus (Linné)+ group		29		4	10	2				1	2		**319**
cf. *L. minutus* (Linné)+													**13**
Cartodere constricta (Gyllenhal)+		1											**4**
Corticaria or *Corticarina* spp.		1											**64**
Total Lathridiidae													**402**
Tenebrionidae													
Tribolium audax Halstead													**1**
T. castaneum (Herbst)+													**2**
T. confusum Jacquelin du Val +													**1**
cf. *Latheticus oryzae* Waterhouse+													**1**
Tenebrio molitor Linné +										1			**23**
T. obscurus Fabricius +													**1**
Neatus tenebriodes (Palisot de Beauvois)													**1**
Total Tenebrionidae													**30**
Meloidae													
Lytta cf. *vesicatoria* Linné+		1											**6**
Salpingidae													
Sphaeriestes virescens (LeConte)		1											**11**
Anthicidae													
Anthicus floralis (Linné)+													**3**
A. formicarius (Goeze)+		3											**21**
Anthicus sp.													**28**
Total Anthicidae													**52**
Bruchidae													
Bruchus sp.+													**1**
Chrysomelidae													
Gastrophysa polygoni (Linné)+													**1**
Hydrothassa vittata (Olivier)*													**1**
Phyllotreta striolata (Fabricius)+													**1**
Crepidodera sp.													**1**
Total Chrysomelidae													**4**
Curculionidae													
Listronotus sp.													**1**

Sub-operation	16D						16E					16K	Total/taxon
Level	**33**	**35**	**38**	**40**	**41**	**42**	**1**	**2**	**3**	**4**	**5**	**5**	
Sitona sp.													**1**
Hypera nigrirostris (Fabricius)+													**3**
cf. *Hypera* sp.													**1**
Curculio sp.		1											**2**
Sitophilus granarius (Linné)+		1				1							**9**
cf. *Sitophilus granarius* (Linné)+													**1**
S. oryzae (Linné)+		3		1	1								**22**
Sitophilus sp.													**1**
Cossonus americanus Buchanan											1		**8**
Dryophthorus americanus Bedel		1	1	1	4	1				7	63	1	**142**
Curculionidae indet. sp.													**1**
Total Curculionidae													**192**
Scolytidae													
Hylurgops rugipennis pinifex (Fitch)		1											**4**
Dendroctinus rufipennis (Kirby)													**2**
D. simplex LeConte		1											**1**
Polygraphus rufipennis (Kirby)		6		1	4								**60**
Pityokteines sparsus (LeConte)													**3**
Orthotomicus caelatus (Eichoff)		1											**2**
Ips sp.													**1**
Dryocoetes affaber (Mannerheim)													**2**
D. autographus (Ratzeburg)													**1**
Crypturgus borealis Swaine													**1**
Xyloterinus politus (Say)													**1**
Trypodendron lineatum (Olivier)													**1**
Xyleborus affinis Eichoff													**1**
Pityophthorus sp.			1										**2**
Scolytidae indet.					1								**4**
Total Scolytidae													**86**
Coleoptera indet. spp.		38	3		10	6				3	1	1	**477**
DIPTERA													
cf. **Ephydridae**		x	x		x							x	
Sphaeroceridae													
Muscoidea		x			x	x							
Muscidae													
Fannia sp.													
Total/level	**9**	**279**	**22**	**36**	**117**	**297**	**2**	**4**	**14**	**30**	**119**	**22**	**6755**

\+ introduced species

* Holarctic species

x present

www.ingramcontent.com/pod-product-compliance
Lightning Source LLC
LaVergne TN
LVHW071215210926
848482LV00015B/813